T0361869

Battery-Integrated Residential Energy Systems

Battery-Integrated Residential Energy Systems introduces battery energy storage systems (BESS) for residential systems and offers insight into modeling, managing, and controlling them. As well as providing a survey of different BESS applications, it explains electrochemical simulation models of BESS. It includes performance parameters, economic analysis, sizing, energy management, control, charging and discharging patterns, coordination of the storage system with other devices in the home, and policy-related matters.

Features:

- Covers different aspects of smart residential energy systems with a battery as a key element.
- Discusses increasing energy efficiency in residential units by optimal control and management of BESS.
- Explores information related to policy regarding the use of BESS.
- Includes chapters on energy trading for implementation of optimal energy management of BESS.
- Focuses on a variety of aspects of battery operation and control.

This book is aimed at senior undergraduate students, graduate students, and researchers in Electrical Engineering, Battery Systems, Energy Engineering, and Sustainable and Renewable Technologies.

Battery-Integrated Residential Energy Systems

Edited by
Asmae Berrada, Altaf Q. H. Badar, and Mohammad Sanjari

CRC Press
Taylor & Francis Group
Boca Raton London New York

CRC Press is an imprint of the
Taylor & Francis Group, an **informa** business

Designed cover image: Shutterstock

First edition published 2025
by CRC Press
2385 NW Executive Center Drive, Suite 320, Boca Raton FL 33431

and by CRC Press
4 Park Square, Milton Park, Abingdon, Oxon, OX14 4RN

CRC Press is an imprint of Taylor & Francis Group, LLC

ISBN: 9781032458762 (hbk)
ISBN: 9781032578385 (pbk)
ISBN: 9781003441236 (ebk)

DOI: 10.1201/9781003441236

Typeset in Times
by Newgen Publishing UK

Contents

Preface

As energy produced from renewable sources is increasingly integrated into residential applications, interest in battery technologies is growing. The field of battery storage systems, their modeling, design, and application has attracted attention recently. Battery energy systems used in residential applications still require more exploration, and they are considered a subject of great interest to the scientific community. This book reviews advances in battery energy storage systems (BESS) for residential systems. The chapters address several aspects of residential BESS, such as modeling, sizing, managing, and controlling them.

The book is divided into thirteen chapters. Chapter 1 introduces battery-integrated residential energy systems while offering some insight into the research that is currently taking place in this area. Chapter 2 looks at different BESS technologies that can be used in residential systems and provides a thorough survey of different BESS applications. Chapter 3 is dedicated to electrochemical simulation models of BESS, which will be of great interest to readers with electrical or chemical engineering backgrounds.

Performance parameters, economic analysis, sizing, and energy management are discussed in the chapters that follow. Control, charging and discharging patterns, and coordination of the storage system with other devices in the home are also discussed. In addition, the book discusses the effects of electricity tariffs on BESS. In designing a residential BESS, these chapters will be particularly useful.

At the moment, BESS are being used passively to supply energy to residential systems. However, they can also be used in an active mode to trade energy, a possibility that is being examined by various researchers. Accordingly, this book contains chapters on decentralized BESS, EV storage systems, community energy storage systems, and the role of BESS in virtual power plants. A BESS's reliability under different conditions is another important factor that is addressed as a separate topic in the book. The last chapter discusses policy-related matters, focusing on the effects of regulatory settings on BESS implementation.

As the power system advances and the grid definitions change to make way for microgrids and nanogrids, the role of BESS will become very important. BESS installed in multiple residential systems will contribute immensely to providing flexibility in the operation of the grid. They will also help reduce losses and improve the reliability and stability of the power systems.

This book will be of interest to energy and power engineers, practitioners in renewable energy storage systems, students, and researchers in energy storage development and implementation. It is hoped that this book will be of value not only to the aforementioned people but also to decision-makers, developers, and all interested people.

Contributors

Youssef Achour
International University
of Rabat, College of
Engineering and Architecture,
LERMA, Sala Al Jadida,
Morocco

Altaf Q. H. Badar
National Institute of Technology
Warangal, India

Asmae Berrada
International University of Rabat,
College of Engineering and
Architecture, LERMA, Sala Al
Jadida, Morocco

Rohit Bhakar
Department of Electrical Engineering,
MNIT Jaipur, Rajasthan

Duong Minh Bui
Department of Electrical and
Computer Engineering, Faculty of
Engineering, Vietnamese-German
University (VGU), Binh Duong
Province, Vietnam

Anisa Emrani
International University of
Rabat, College of Engineering
and Architecture, LERMA Lab,
Sala Al Jadida, Morocco

Naoufel Ennemiri
International University of Rabat,
College of Engineering and
Architecture, LERMA Lab, Sala Al
Jadida, Morocco

Mohammad Reza Fallahzadeh
Electrical Engineering Department,
Shahid Rajaee Teacher Training
University, Tehran, Iran

Shivani Garg
Department of Electrical Engineering,
MNIT Jaipur, Rajasthan

Ikram El Haji
National School of Mines Rabat,
Morocco

Rohini Haridas
Department of Electrical Engineering,
MNIT Jaipur, Rajasthan

Abdennebi El Hasnaoui
National School of Mines Rabat, Morocco

Truong Hoang Bao Huy
Department of Future Convergence
Technology, Soonchunhyang
University, Asan-si, Chuncheongnam-
do, South Korea

Bhavna Jangid
Centre of Energy & Environment,
MNIT Jaipur, Rajasthan

Mustapha Kchikach
National School of Mines Rabat,
Morocco

Daehee Kim
Department of Future Convergence
Technology, Soonchunhyang
University, Asan-si, Chuncheongnam-
do, South Korea

Phuc Duy Le
Administration Department, Industrial
    University of Ho Chi Minh City,
    Vietnam

Wei Liu
School of Electrical and Electronic
    Engineering, Universiti Sains
    Malaysia, Nibong Tebal, Penang,
    Malaysia; College of Mechanical and
    Electrical Engineering, Cangzhou
    Normal University, China

Parul Mathuria
Department of Electrical Engineering,
    MNIT Jaipur, Rajasthan

Asmae El Moukrini
Rabat Business School, International
    University of Rabat, BEAR LAB,
    Salé, Morocco

Arun Nayak
Department of Electrical Engineering,
    MNIT Jaipur, Rajasthan

Debasmita Panda
National Institute of Technology
    Warangal, India

Chandra Prakash
Department of Electrical Engineering,
    MNIT Jaipur, Rajasthan

Sanaa Sahbani
National School of Mines Rabat, Morocco

Mohammad Sanjari
School of Engineering and Built
    Environment, Griffith University,
    QLD, Australia

Ali Shayegan-Rad
MAPNA Electric and Control,
    Engineering and Manufacturing
    Co. (MECO), MAPNA GROUP,
    Alborz, Iran

Jian Shi
School of Electrical and Electronic
    Engineering, Universiti Sains
    Malaysia, Nibong Tebal, Penang,
    Malaysia

Swasti Swadha
Department of Electrical Engineering,
    MNIT Jaipur, Rajasthan

Jiashen Teh
School of Electrical and Electronic
    Engineering, Universiti Sains
    Malaysia, Nibong Tebal, Penang,
    Malaysia

Khoa Hoang Truong
Department of Power Delivery, Ho Chi
    Minh City University of Technology
    (HCMUT), Vietnam; Vietnam
    National University Ho Chi Minh
    City, Vietnam

Ajay Verma
Department of Electrical Engineering,
    MNIT Jaipur, Rajasthan

Dieu Ngoc Vo
Department of Power Systems,
    Ho Chi Minh City University of
    Technology (HCMUT),
    Vietnam; Vietnam National
    University Ho Chi Minh City,
    Vietnam

Cyrus Wekesa
University of Eldoret, Kenya

Shubham Yadav
Department of Electrical Engineering,
  MNIT Jaipur, Rajasthan

Ali Zangeneh
Electrical Engineering Department,
  Shahid Rajaee Teacher Training
University, Tehran, Iran

About the Editors

Asmae Berrada is an associate professor of Energy in the school of Energy Engineering at the international University of Rabat (UIR). She holds a PhD degree from the Faculty of Science USMBA. Her PhD work was prepared at Al Akhawayn University within the framework of the EUROSUNMED project (EU-H2020). She obtained a Master's degree in Sustainable Energy Management, and a Bachelor's degree in Engineering and Management Science from Al Akhawayn University. Dr. Berrada started her career as a part-time faculty member at Al Akhawayn University from 2014 to 2018. She has been ranked among the World's Top 2% Scientists in the Stanford University and Elsevier ranking for 2023. She has also received several research awards and honors from IRESEN, MASEN, and USMBA. She has led a number of research projects on renewable energy, hybrid energy systems, and energy storage systems such as SHPS, GESYS, and DARNASOL. She has published a variety of articles in leading international energy journals, and two books (*Gravity Energy Storage* and *Hybrid Energy System Models*). Her main research focus is on energy systems and their modeling. She has actively been working in the fields of energy storage and renewable energy.

Altaf Q. H. Badar is currently associated with the Electrical Engineering Department of National Institute of Technology Warangal, India. He completed his Bachelor's and Master's degrees at Rashtrasant Tukdoji Maharaj Nagpur University, India in 2001 and 2009, respectively. He pursued his PhD at Visvesvaraya National Institute of Technology, India and completed it in 2015. His PhD research dealt with reduction of active power losses by controlling the flow of reactive power in the transmission system through evolutionary optimization techniques. He also has a Diploma in Business Management completed in 2004 from ICFAI, India. He has approximately 17 years of experience of teaching professional engineering courses and four years of field experience. His research interests include evolutionary optimization techniques, energy management, smart homes, and energy trading. He has published his research work in more than 30 international papers in journals and conferences. He was awarded a project under the RSM scheme of his parent institution on the topic of smart homes. He has also executed a project awarded under GIAN worth $12,000 on the topic of Advanced Energy Conversion and Storage Systems. He has been a member of the board of studies of different institutions, including RTM Nagpur University, India. He has also chaired a number of international conferences and been a member of various technical committees. He was an editorial member of *International Refereed Journal of Engineering Science and Technology* and is currently an editorial member of *American Journal of Electrical Power and Energy Systems* and *International Advanced Research Journal of Engineering and Technology*. He has to his name a book on *Evolutionary Optimization Algorithms* and around five book chapters. He has delivered his services as a reviewer for multiple international journals and conferences and for a book. He has delivered a large number of expert lectures in different engineering institutions across India. He has guided around 30 undergraduate and ten postgraduate projects. Currently, 40 research scholars are working under his guidance. He is a Senior Member of IEEE, Life Member of the Indian Society of Technical Education, and a member of IAENG. He understands seven different languages at various levels.

Mohammad Sanjari (PhD, FHEA) received his PhD in Electrical Engineering in 2013 after successfully accomplishing the project titled "Control of Microgrid using Distributed Object-Oriented Intelligent System with Partial Information". In this project he investigated a partial information- based distributed control scheme for a microgrid aiming to enhance reliability of renewable energy-integrated power system across different modes of operation while addressing cyber system challenges. Currently, he is affiliated with the School of Engineering and Built Environment at Griffith University, Australia. As an accomplished researcher, he has authored over 90 journal and conference papers. His fields of expertise are applying artificial intelligence and data mining in low carbon power systems analysis and control. He has led several research projects focused on decision making in power systems, considering network security and risk constraints by using machine learning methods and big data analysis techniques. He also has expertise in probabilistic analysis of smart multi-carrier energy systems, including hybrid load demands and renewable energy resources.

1 Introduction to Battery-Integrated Residential Energy Systems

Asmae Berrada, Altaf Q. H. Badar, and Mohammad Sanjari

1.1 INTRODUCTION

The operations of the grid have experienced numerous transformations, driven by the continuous emergence of novel technologies. The consumer's habits and level of power consumption have changed, and their activities have also changed because of the varying lifestyle. The level and pattern of a consumer's power consumption typically correlates with the level of development in the country where the consumer resides.

An important transformation involves incorporating renewable energy sources (RES) into the grid. RES can be installed on a large scale on the generation side or in smaller units, such as solar rooftops, which can be installed on the consumer side. The most widely used RES on the distribution side are solar and wind. Residents can utilize the energy generated from these resources to supply their own demand or sell it to the grid. Consumers who install RES on their premises and actively utilize resources for energy trading are termed prosumers.

RES, like solar and wind turbines, generate electricity using solar radiation and wind flow. However, these resources are not continuously available. Solar irradiation is available during the day, and the wind may or may not flow at a given time. This intermittency plays a very important role in the usage of these RES and is the major demerit in their usage. To overcome this demerit, battery energy storage systems (BESS) are implemented. BESS store energy when there is excess generation, that is, when the energy generated from RES is more than the demand, and supply energy when there is deficit energy. BESS thus provides stability and increases the reliability of the system. BESS can also be installed in such a way as to supply energy for long periods of time.

In countries where the grid is deregulated, the applied electricity tariff generally varies with time, for example, a time of use (TOU) tariff. In the TOU tariff system, energy prices vary during the day, depending on the load-generation imbalance. BESS can be actively implemented to reduce energy bills by charging the batteries when the tariff is low and discharging the batteries when the tariff is high.

Smart home energy management systems (SHEMS) play a vital role in maximizing the use of BESS. SHEMS are able to communicate with different

DOI: 10.1201/9781003441236-1

appliances in the home, BESS, user, and utility. SHEMS are able to forecast energy prices and hence optimize the schedule of different appliances and of BESS. The state of charge (SOC) of BESS is taken into consideration while optimizing its charging and discharging cycles. In addition, the charging and discharging efficiency as well as the charging and discharging rates are utilized in the evaluation of the SOC.

In any home, there are different kinds of loads, such as shiftable and non-shiftable, continuous and non-continuous. Shiftable loads are those appliances that can be turned on or off at different times. The time of usage of these appliances is not fixed. Non-shiftable loads, on the other hand, cannot be shifted and have to be utilized at the given time itself. An example of a shiftable and a non-shiftable load would be a water pump and lights, respectively. Continuous loads are those loads that have to be utilized in a single run and cannot be turned on and off in a cycle. Non-continuous loads are those loads whose utilization can be broken. An example of a continuous and a non-continuous load would be an oven and a washing machine, respectively. The presence of BESS in the home provides more flexibility in scheduling the loads during the day based on various characteristics, factors, and constraints. BESS can be used to supply power during peak periods and also in the absence of power supply. The reliability of the residential system is therefore enhanced.

The presence of BESS in the distribution network not only increases reliability and flexibility but also improves the stability of the system. BESS can be utilized to help introduce resilience to the grid and plays a crucial role in improving the overall performance, operation, and control of the grid.

1.2 EMERGENCE OF BATTERY-INTEGRATED RESIDENTIAL SYSTEMS

The initial references to battery-integrated residential systems appeared around 1976. Among the literature citing the usage of BESS in a household is [1]. This paper presents an overview of the usage of BESS in a household, and the charging and discharging levels to which the batteries can be subjected.

In 1977, a report on solar photovoltaics in the context of the residential consumer was presented [2]. It covers around seven regions of the USA in its analysis. Battery design in relation to its construction is also investigated. An optimal battery size evolved from the studies.

A report presented in 1978 related to the charging of BESS from multiple sources and photovoltaics alone [3]. It introduces solar generation forecasting. Three geographical regions are considered: mountain, plane, and coastal. A review of eleven storage systems is conducted.

A residential-size photovoltaic system is studied in [4]. The tests were conducted at MIT Lincoln Laboratory's Photovoltaic System Test Facility in Concord, Massachusetts. A controller for residential photovoltaic systems is presented. The study considers different residential loads.

The research publications have explored different aspects of integrating batteries with residential systems. Some of the important objectives are listed below:

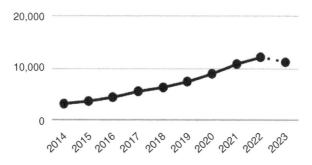

FIGURE 1.1   Number of publications having keywords *residential* and *battery*.

a. Sizing of batteries [5];
b. Economic optimization [6];
c. Economic analysis and control strategy [7];
d. Degradation analysis of BESS [8];
e. Scheduling of BESS [9];
f. Techno-economic analysis [10, 11];
g. Energy trading [12, 13]; and
h. Demand response [14].

The number of publications having keywords *residential* and *battery* in the Engineering and Electrical Engineering category is plotted in Figure 1.1 [15].

1.3   RESIDENTIAL ENERGY ISSUES

Environmental awareness, continuous availability of power, and energy prices are some of the major factors that bother today's residential consumers. In this section, the role of BESS in mitigating the adverse effects of the above factors is highlighted.

Energy generation is still heavily dependent on conventional fossil fuels. According to [16], the energy mix in the generation has witnessed a lot of changes. The changes in the energy mix are positive and have been realized because of the policies being adopted by the governments of different countries in the world. In 2013, around 40 percent of total energy was generated through coal-fired power plants, which was reduced to around 35 percent in 2022 (Figure 1.2). On the other hand, energy generation from wind and solar has seen a significant jump in percentage during the same period. The total energy generation in 2013 was 10.52 million GWh, which increased to 10.84 million GWh in 2022 [17].

There are two aspects through which energy generation can be made greener and more environmentally friendly. The first aspect is to produce more energy through the use of RES such as wind and solar. The second aspect consists of managing, optimizing, reducing, and controlling energy demand. Residential load accounts for around 30 percent of the total load. An efficient scheduling and utilization of the latter can be very effective in achieving the environmental goals. A residential load cannot be effectively scheduled without the presence of BESS.

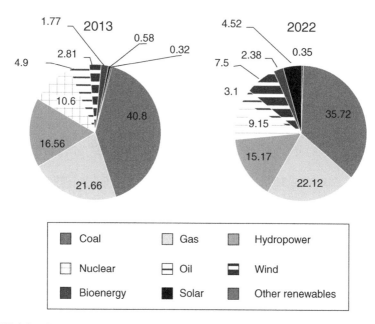

FIGURE 1.2   Comparison between energy generation in 2013 and 2022.

In some developing and underdeveloped countries, the availability of continuous power is still a distant dream [18]. Natural events such as tornadoes, earthquakes, snow, and rain can cause disruptions in power supply. In February 2021, severe winter storms were the cause of a major power outage in Texas. The unavailability of power is a cause of great discomfort.

BESS can be installed to provide power during such times. Residential consumers have turned to the installation of RES, like solar and wind, on their premises for the better. However, these sources are not reliable and require BESS as a buffer to store excess energy and provide energy in times of deficiency. BESS helps convert a residential place into a nanogrid that is self-sufficient when combined with renewable energy technologies.

Energy costs for consumers are calculated through different methods. The tariff system varies from country to country. Block rate tariffs and TOU tariffing are some of the most common methods of billing energy. For block rate tariffs, the per unit cost of energy is constant for a certain number of units. The per unit cost increases after these limits are crossed and remains constant until the next block limit. On the other hand, in TOU, the energy tariff changes with time and is usually dependent on the imbalance between generation and consumption. The presence of BESS in residential systems helps in optimizing the scheduling of loads. The loads are turned on when the prices are low, and vice versa. However, BESS can supply electricity for non-shiftable loads utilized during peak tariff hours, thus lowering the overall cost of energy.

1.4 REGULATORY AND POLICY ASPECTS

The adoption of battery-integrated residential technologies is significantly influenced by policies and regulations. This landscape of such cutting-edge solutions is being shaped by governments and regulators, which is very important as RES and energy storage technologies are increasingly necessary in order to secure an efficient and sustainable energy future.

These policies and regulations will be great drivers for the rapid pace at which battery-integrated residential energy system solutions are deployed. Financial incentives, standards, and policies may enable a smoother shift to more resistant and greener energy systems. This would require amendments to the rules and regulations by the regulators and governments to promote battery-integrated solutions as an alternative, given the fact that the energy landscape is rapidly changing with the transition from fossil fuels to greener resources. Increasing the market for such systems based on a demand-driven approach is often considered to be one of the major aims when supporting battery-integrated residential solutions [19]. For example, the feed-in tariff for renewable energy generation has directly and indirectly influenced the development of battery storage systems in Germany [20]. Structural changes in legislation and regulations in the form of financial incentives and subsidies are amongst the most effective means of influencing acceptance in favor of battery-integrated solutions. Few governments across the world have started elaborating programs that extend financial support to specific residentials that are willing to adopt energy storage devices. Such inducements include monetary grants, low-interest loans, tax credits, and refunds.

Another key regulatory feature is the ability to integrate battery-integrated systems into the electric grid and participate in net metering programs. Under net metering plans, households can get credit on their utility bills if they feed surplus power from renewable sources, such as solar, into the grid. Clear and attractive net metering regulations are essential if they encourage residential customers to sell energy. This will increase the economic attractiveness of investing in those types of systems. The technical and safety requirements for integrating battery systems into the grid are set by regulatory organizations. Some requirements are needed for household energy systems to work safely and smoothly. Utilities are major players in the energy environment. The treatment of battery-integrated systems under regulatory frameworks covering homes and utilities may help or hinder the adoption process. To create such an environment, one needs to formulate clearly defined rules for grid connection, faster review procedures, and equitable compensation for any excess energy sent back to the system.

Environmental regulations may also affect the adoption of battery-integrated technologies. These rules may have some impact on the manufacturing and disposal of these systems. More environmentally friendly battery technology and recycling procedures could be developed in response to environmental regulations. Governments that have energy savings, carbon reduction, and renewable energy goals often set targets for the deployment of energy storage. These goals could provide direction and a sense of urgency for the industry.

As these are the goals set by regulators, the market will still see demand for battery-integrated residential solutions. These regulations are very helpful in areas that suffer from severe weather or weak grids. They recognize the role of storage and distributed energy resources in maintaining electrical supplies during shortages. In order to promote the use of battery-integrated systems, consideration should be given to improving policies. The following are some suggestions for redesigning deployment rules to encourage innovation, making the regulatory laws applicable throughout the battery life cycle, and strengthening policies' comprehensiveness [19–23].

1.5   ENVIRONMENTAL AND SUSTAINABILITY ASPECTS

The shift to greener, more sustainable energy development must include battery-integrated residential systems. These systems are not only economically sensible for households; they also represent a first step towards reducing the environmental risk associated with fossil fuel sources. They reduce greenhouse gas emissions and help stabilize grid networks while at the same time promoting RES. As long as policy and technology continue to support these systems, one can anticipate that their contribution to a more environmentally friendly future will increase even further.

The incorporation of battery-integrated solutions into residential energy systems has a huge impact on sustainability and the environment. Such energy storage systems, coupled with renewable energy systems, are the key to reducing greenhouse gases and building a more friendly environment. One of the foremost environmental advantages offered by battery-integrated systems is reducing greenhouse gas emissions. These systems help reduce the reliance on fossil fuels by storing excess energy and discharging it during periods of low production or high demand. As a result, power-generating carbon emissions are reduced. This is a cleaner and greener energy source which helps lower global warming [24, 25].

When coupled with intermittent RES like solar and wind, battery-integrated residential systems greatly enhance the stability of the electrical grid. As mentioned before, these technologies can store surplus energy when production exceeds demand and release it when needed. This helps reduce the need for backup (fossil fuel) power plants, which would otherwise have to be used in case there are fluctuations in demand, and therefore further reduces greenhouse gas emissions as well as air pollution. Adding intermittent RES with battery integration can make electrical grids more stable and less subject to failure or blackouts. As fewer backup fossil fuel power units are needed to cope with demand fluctuations, greenhouse gas emissions and air pollution can be further reduced.

Another important point to be considered is how sustainable energy storage technology is. As the industry matures, a more environmentally friendly battery system and recycling process are playing an increasingly important role. The change to lithium-ion phosphate batteries is crucial, as they are well known for their strength and safety but low environmental impact [26].

Battery-integrated residential systems successfully fulfill two roles. By using the excess energy in batteries to charge electric vehicles (EVs), residents can reduce their dependence on gasoline-powered cars, cutting further greenhouse gases and improving air quality. In addition, battery-integrated residential systems can also reduce peak load. This means that utilities could avoid or delay costly infrastructure expansion projects. Adopting these solutions lessens the environmental impact of developing and operating new power plants in addition to saving money.

1.6 ECONOMIC CONSIDERATIONS

The decision to install battery-integrated systems can be influenced by a number of economic considerations, including cost-effectiveness, return on investment (ROI), and financial incentives. Governments, utilities, and other organizations must provide financial incentives in order to encourage the use of battery-integrated residential energy systems. Among these incentives are low-interest loans, tax credits, grants, and refunds. For example, some residents might receive a tax credit equal to part of the price they pay for an energy storage system. This would reduce their initial expenditure outlay. Financial incentives can make these solutions more attractive and accessible by removing the financial barrier at some level. The effective use of power boosts power systems' financial sustainability [27]. Rising solar feed-in rates are the impetus for the widespread adoption of battery residential systems [28].

One financial factor residential customers consider when deciding whether to install battery-integrated systems is the return on investment (ROI). This is expressed as the ratio of the initial cost to the economic benefits (lower energy bills, and revenue from grid services). Therefore, a shorter ROI period is usually attractive. Battery systems with a faster return are often chosen because they can help households save and recover investment faster, particularly when paired with solar panels [29]. For example, a battery energy storage system has an internal rate of return as high as 13.14 percent in a case study carried out in the UK [30].

The cost-saving potential for installing domestic energy storage systems is one of the main factors. Battery systems, however, can usefully reduce the need to buy power from the grid by storing up excess energy generated during low-demand time intervals and releasing it during high-demand periods. Among the factors that determine how much can be saved are, in particular, the size of battery system, power tariffs, and patterns of energy use [31]. Battery-integrated systems are allowed to participate in grid services and receive TOU prices. Grid services enable battery owners to earn money by providing grid stability solutions, while revised TOU tariffs give residential customers incentives to consume power outside of peak periods. Both of these elements can increase the financial advantages to energy storage devices.

When combined with a solar system, battery-integrated systems can also optimize financial returns. Moreover, since solar electricity can be stored in the battery and used at night when there is no sunshine, it becomes unnecessary to buy power from the grid. Synergy can generate large savings and early returns on investment, both of which boost the financial attractiveness of investing in these systems [29, 32]. With the emergence of options for low-interest loans or leasing programs, battery-integrated systems may become cheaper and more attractive. These choices allow homeowners to spread out their investments and reduce their spending. Additionally, battery-integrated systems can offer security and energy independence in the event of blackouts. This is especially tempting for residential customers in areas where grid outages are frequent. So these systems do have some economic value since, in the event of an outage, they can still provide basic services and appliances.

As far as battery-integrated home energy systems are concerned, everything depends on finance. A number of factors affect the decision to invest in these systems, such as financing options, grid services, and solar integration potentials. As long as technology continues to progress and costs continue to decline, battery-integrated systems will be an attractive option for residential customers who want grid stability, sustainability, and cost savings.

1.7   CHALLENGES AND LIMITATIONS OF RESIDENTIAL BATTERY ENERGY SYSTEMS

Integrated batteries have their advantages in residential energy systems, but users should understand that they also come with disadvantages and restrictions. These include initial cost, operation and maintenance (O&M), and system limitations. The high initial cost of buying and installing a battery-integrated system is the biggest challenge. Although the cost of batteries has fallen recently, it still accounts for a major portion of overall system costs. This remains a hurdle for many homeowners, despite available financing and incentives. Although battery-integrated systems usually require very little maintenance, they may be some expenses, particularly as the system becomes more mature. Sometimes maintenance and monitoring are required to ensure peak performance. In addition, replacing a battery can be expensive, and it has to be done after four or five years [33].

Another problem is the fluctuation of the system's ROI. The size of the system, local electricity prices, and patterns of energy consumption are just a few factors that influence battery-integrated systems' ROI. However, getting a positive ROI is difficult, especially in locations with cheap energy prices or restrictive TOU tariff regimes. In addition, a battery has a limited lifetime, usually measured in cycles or years. They need replacements from time to time. How long a battery lasts depends on temperature, drain depth, and use habits. Furthermore, the efficiency and capacity of battery technology are limited. Currently, the most widely used batteries for home systems are lithium-ion batteries. Their energy capacity is limited, and efficiency drops during the charging and discharging modes. Hence, they may not be the best choice in every case or able to meet all of a household's energy needs.

The installation of battery-integrated systems requires physical space, which is not always available. The battery itself needs enough room to be safe and last for a long time; therefore, there must also be space for ventilation and temperature control. Even though they produce relatively low greenhouse gas emissions, manufacturing, shipping, and eventual disposal have an environmental impact. Therefore, proper disposal and recycling procedures are important in reducing this impact.

Battery-integrated systems lower residential customers' electricity bills, but standards governing connectivity and variations in regional regulatory laws may make it difficult for residential customers to install the new system. Utilities may also place limits on the amount of kWh or watt hours that can be generated. Occasionally, administrative obstacles or imprecise directions can slow down the adoption process. There are also compatibility and technological development issues. Advances in battery technology may make modern systems out of date rapidly. Finding newer, better batteries that work with existing systems could be complex for users.

Given these challenges and limitations, it is worth noting that the industry continues to develop; many of these problems are now being tackled through ongoing research and development work. As battery system costs continue to decrease, the use of battery-integrated residential energy systems is expected to increase, making them more attractive for residential customers looking for energy cost reduction and resilience.

1.8 MARKET TRENDS

A market for battery-integrated residential has been created because of the technological progress, the changing user preferences, and the increasing awareness about the benefits these systems can provide. This market is expanding rapidly in recent years and has shown an increase in demand due to the growing need for flexibility and self-consumption. In addition, many reasons have influenced this rise, including falling battery prices and environmental concerns related to the current energy production. Many residentials have combined solar panels with battery systems to fully use renewable energy resources and reduce their dependence on the traditional electric grid. Installation of battery systems is often boosted by TOU tariffs offered by power providers. At these rates, households are encouraged to use less energy during off-peak periods and minimize their consumption during peak hours. Incentives that promote the adoption of battery-integrated residential systems include tax credits and refunds [34].

Power outages have triggered a high interest in battery-integrated systems. Residential battery systems make sure that residents have electricity when the power goes off. Apart from this, residents have become more and more eager to produce their own electricity for consumption. Residentials with integrated battery systems have greater control over the production and use of electricity, lessening dependence on outside power sources. As environmental awareness grows, customers are increasingly interested in reducing their carbon footprint and are seeking sustainable green energy solutions. These sustainability goals can be achieved through the use of cleaner energy sources and battery-integrated systems. With the market's growth, manufacturers and service providers face greater competition. This competition results in innovation and cost-cutting that consumers benefit from. Moreover, every new development in battery technology makes energy storage systems more reliable, durable, and safer. For example, lithium-ion batteries are more stable and have better performance.

Many countries have enacted laws and legislation promoting the use of battery-integrated residential energy systems. Some of these policies include financial rewards, net metering initiatives, and goals for the uptake of renewable energy. The market for integrated battery energy systems is expected to continue growing steadily in the future, according to forecasts by market analysts. As technology advances and costs fall, the adoption rate should increase. This growing trend is pushing forward the development of household energy systems with integrated batteries. EVs and renewable energy (RE) make use of batteries or battery storage facilities built for the automobile industry to offer support services such as momentary relief power at times when grid voltage has become unstable [35]. By 2050, world-wide peak EV battery capacity is expected to reach between 32 and 62 TWh.

Li-ion batteries offer demand response (DR) and residential flexibility. By adjusting the load of household appliances, including electric cars, residential end users can add value to the power system and reduce overall system costs by about 1 percent. According to [36], it seems that demand for residential energy systems with integrated batteries is being driven by the many advantages offered by load shifting and DR programs. In Europe, the market penetration of decentralized battery systems for home self-consumption is expected to reach 467 GWh by 2050. A number of factors, including growing battery installations, technological costs, and future energy prices, are driving the market adoption of decentralized battery systems for residential self-consumption [37]. This demonstrates the expanding market potential for integrated battery-powered residential energy systems.

To summarize, there is now increased demand for and innovation in residential energy systems utilizing integrated batteries. Declining battery prices, environmental awareness, grid resilience, and government incentives are all driving the adoption of batteries. As long as users keep looking for ways to become less dependent and decrease their carbon emissions, the market will continue expanding and changing, and of course, technology will continue developing.

1.9 FUTURE PROSPECTS OF BATTERY-INTEGRATED RESIDENTIAL SYSTEMS

The future of battery-integrated household energy systems will be radically transformed through innovations prompted by market forces, continuing technical advances, and further research. For battery-integrated home energy systems, the following are some significant future developments, emerging technologies, and research avenues to consider.

One of the main objectives is to further develop battery technology. Researchers are continuously improving the energy density, safety, and performance of batteries. Next-generation battery systems, such as solid-state batteries, should deliver greater energy storage capacity, a longer lifespan, and enhanced safety. In addition, extending the life of batteries is an important objective. Scientists are working on slowing down the deterioration of battery cells over the life cycle, which should reduce maintenance costs and give longer system lifetimes. By improving the energy density and efficiency of batteries, it is possible to store more energy in a smaller amount of space and minimize energy losses during the battery charge and discharge cycles.

Another objective of R&D is to improve the way household battery systems are connected to the power grid. Although still a relatively new concept, the idea of virtual power plants (VPPs) comprises networks of household energy systems that work together to ensure grid stability and reduce peak demand for electricity. Another new trend is the use of systems, including energy storage, with EVs or other power technologies such as heat pumps for heating and cooling. With these hybrid systems, homeowners can now receive more complete and effective energy solutions than ever before [38]. In addition, artificial intelligence (AI) and machine learning are making battery-integrated energy systems increasingly effective. Complex algorithms can be used to guide and direct energy flows, maximizing savings and efficiency [39]. In the energy sector, blockchain technology is being studied as a means to enable peer-to-peer trading. It promotes energy independence and decentralization, as homes can buy or sell excess power directly to other customers.

Research into environmentally friendly battery components and recycling methods is also necessary. Therefore, particular attention should be given to developing solutions that reduce waste and the environmental impact of battery production and disposal. As climate change-related disasters and power outages become increasingly frequent, making homes more energy-secure is becoming necessary. Improving backup power capacities and integrating battery systems with other onsite generating sources, such as microgrids, are the two main areas of research.

The energy regulatory environment is expected to change in order to better accommodate battery-integrated home energy systems. These measures include the creation of uniform interconnection procedures, fair compensation for grid services, and rules that encourage batteries to provide tracking services to the grids. Finally, measures are now under way to reduce costs so that the price of battery-integrated systems can be reduced even further for a wider market. An important objective is to reach cost parity with existing energy sources. If customers are to use battery-integrated systems, they must first become familiar with their benefits and features. For households, there will be tools for data analysis and easy-to-use interfaces so that their energy consumption can be more consciously managed.

In terms of prospects for a low-carbon future, battery residential systems also offer good opportunities [40]. These devices can provide the electrical grid with many functions, including peak shaving and frequency control [41]. When batteries are used for second-life applications, integrating them into home systems would be the best way to avoid excessive waste of resources and expenses [42]. Battery storage use needs to be expanded through a restructuring of regulations and a change in market principles [43]. In general, home systems that use batteries have the capability to facilitate the shift towards a more robust and sustainable energy system [44].

Finally, battery-integrated home energy systems are full of potential for development and expansion. These systems will also be improved by future research and technical advances, as well as the development of energy management technologies. Battery-integrated residential energy systems will perhaps play an important role in the development of greener and more robust energy systems as they become stronger, cheaper, and more effective.

1.10 CONCLUSION

Economic, policy, and regulatory factors will all affect the use of battery-integrated residential systems. But for these types of designs to be used in residential applications, financial incentives and a clearly defined legal system are required. Besides, from an environmental perspective, these systems help promote self-sufficiency and reduce levels of greenhouse gases. The economic benefits of these systems, particularly the ROI and cost-effectiveness factors, greatly influence residential costumers' decisions; as they enable them to both save money and improve grid stability. This makes battery-integrated residential systems particularly attractive as a solution.

The prospects for such integral home battery systems look good. Ongoing research in battery technology will improve energy density, safety, and performance. That is why concepts such as VPPs and hybrid systems that pair energy storage with electric car charging are gradually becoming more integrated or economical. These systems become more decentralized and operationally accountable when combining blockchain, machine learning, and AI.

These battery-integrated home energy systems could be the key to a sustainable, low-carbon future. Technology has advanced, market trends have changed, and a suitable regulatory environment has made these systems an integral part of the green transition. It is believed that as batteries provide longer life, greater affordability, and time effectiveness, these systems will have a greater impact on the environment.

REFERENCES

[1]   Miller, D. B., & Boeer, K. W. (1976). Solar-electric residential system tests. *Energy Development II. (A78-10729 01-44)*. Institute of Electrical and Electronics Engineers, New York, 116–124.

[2]   Feduska, W., et al. (1977). *Energy storage for photovoltaic conversion. Volume III. Residential systems.* Final report No. TID-28779. Westinghouse Research and Development Center, Pittsburgh, PA (USA).

[3]   U.S. Department of Energy. (1978). *Applied research on energy storage and conversion for photovoltaic and wind energy systems. Volume II. Photovoltaic systems with energy storage.* National Technical Information Service, Springfield, VA. www.osti. gov/servlets/purl/7071934/

[4]   Cadieux, R. N. (1979). *Battery charge/solar array controller for a residential size photovoltaic power system.* No. COO-4094-62. Massachusetts Inst. of Technology (MIT), Lincoln Lab, Lexington, MA (USA).

[5]   Weniger, J., Tjaden, T., & Quaschning, V. (2014). Sizing of residential PV battery systems. *Energy Procedia, 46,* 78–87.

[6]   Hesse, Holger C., et al. (2017). Economic optimization of component sizing for residential battery storage systems. *Energies, 10*(7), 835.

[7]   Badar, Altaf Q. H., & Sanjari, M. J. (2020). Economic analysis and control strategy of residential prosumer. *International Transactions on Electrical Energy Systems, 30*(9), e12520.

[8]   Mishra, P. P., et al. (2020). Analysis of degradation in residential battery energy storage systems for rate-based use-cases. *Applied Energy, 264,* 114632.

[9]   Ratnam, E. L., Weller, S. R., & Kellett, C. M. (2015). An optimization-based approach to scheduling residential battery storage with solar PV: Assessing customer benefit. *Renewable Energy, 75,* 123–134.

[10]   Goebel, C., Cheng, V., & Jacobsen, H.-A. (2017). Profitability of residential battery energy storage combined with solar photovoltaics. *Energies, 10*(7), 976.

[11] Barcellona, S., et al. (2018). Economic viability for residential battery storage systems in grid-connected PV plants. *IET Renewable Power Generation,* 12(2), 135–142.

[12] Guo, Z., et al. (2020). Optimal energy management of a residential prosumer: A robust data-driven dynamic programming approach. *IEEE Systems Journal,* 16(1), 1548–1557.

[13] Dimitroulis, P., & Alamaniotis, M. (2022). A fuzzy logic energy management system of on-grid electrical system for residential prosumers. *Electric Power Systems Research,* 202, 107621.

[14] Leadbetter, J., & Swan, L. (2012). Battery storage system for residential electricity peak demand shaving. *Energy and Buildings,* 55, 685–692.

[15] Dimensions. Retrieved from https://app.dimensions.ai/

[16] Ritchie, H. (2023, October 9). Electricity Mix. Retrieved from https://ourworldind ata.org/electricity-mix#:~:text=Globally%20we%20see%20that%20coal,drama tic%20changes%20in%20over%20time

[17] Energy – Electricity Generation – OECD Data. (n.d.). Retrieved from https://data. oecd.org/energy/electricity-generation.htm

[18] Gubangxa, K. (2023, December 20). Major Countries with Load Shedding. Retrieved from https://elum-energy.com/blog/major-countries-with-load-shedding/

[19] Wang, X., Huang, L., Daim, T., Li, X., & Li, Z. (2021). Evaluation of China's new energy vehicle policy texts with quantitative and qualitative analysis. *Technology in Society,* 67, 101770. https://doi.org/10.1016/j.techsoc.2021.101770

[20] Sinsel, S. R., Markard, J., & Hoffmann, V. H. (2020). How deployment policies affect innovation in complementary technologies—evidence from the German energy transition. *Technological Forecasting and Social Change,* 161, 120274. https://doi.org/ 10.1016/j.techfore.2020.120274

[21] Barkhausen, R., Fick, K., Durand, A., & Rohde, C. (2023). Analysing policy change towards the circular economy at the example of EU battery legislation. *Renewable and Sustainable Energy Reviews,* 186, 113665. https://doi.org/10.1016/ j.rser.2023.113665

[22] Park, C., & Kim, J. K. (2021). Study on policy tasks for promoting a business using spent electric vehicle batteries. *Designs,* 5(1), 14. https://doi.org/10.3390/designs 5010014

[23] Berrada, A., & Loudiyi, K. (2019). *Gravity Energy Storage.* Elsevier, Netherlands.

[24] Abdelkareem, M. A., Ayoub, M., Khuri, S., Alami, A. H., Sayed, E. T., Deepa, T. D., & Olabi, A. (2023). Environmental aspects of batteries. *Sustainable Horizons,* 8, 100074. https://doi.org/10.1016/j.horiz.2023.100074

[25] Cusenza, M. A., Guarino, F., Longo, S., Ferraro, M., & Cellura, M. (2019). Energy and environmental benefits of circular economy strategies: The case study of reusing used batteries from electric vehicles. *Journal of Energy Storage,* 25, 100845. https:// doi.org/10.1016/j.est.2019.100845

[26] Coban, H. H. (2022). Production and use of electric vehicle batteries. *Energy Systems Design for Low-Power Computing,* edited by R. R. Gatti et al., 279–304. IGI Global, Hershey, PA. https://doi.org/10.4018/978-1-6684-4974-5.ch014

[27] Dagnachew, A. G., Choi, S. M., & Falchetta, G. (2023). Energy planning in Sub-Saharan African countries needs to explicitly consider productive uses of electricity. *Scientific Reports,* 13(1), 13007. https://doi.org/10.1038/s41 598-023-40021-y

[28]   Best, R., Li, H., Trück, S., & Truong, C. (2021). Actual uptake of home batteries: The key roles of capital and policy. *Energy Policy*, 151, 112186. https://doi.org/10.1016/j.enpol.2021.112186

[29]   Aniello, G., Shamon, H., & Kuckshinrichs, W. (2021). Micro-economic assessment of residential PV and battery systems: The underrated role of financial and fiscal aspects. *Applied Energy*, 281, 115667. https://doi.org/10.1016/j.apenergy.2020.115667

[30]   Pereira, C. O., Torquato, R., Freitas, W., & Ding, H. (2023). Wide-scale assessment of the payback of a battery energy storage system connected to MV customers. *IEEE Transactions on Sustainable Energy*, 14(3), 1909–1912. https://doi.org/10.1109/tste.2023.3235213

[31]   Loudiyi, K., & Berrada, A. (2014). Operation optimization and economic assessment of energy storage. *2014 International Renewable and Sustainable Energy Conference (IRSEC)*. https://doi.org/10.1109/irsec.2014.7059828

[32]   McIlwaine, N., Foley, A. M., Kez, D. A., Best, R., Lu, X., & Zhang, C. (2022). A market assessment of distributed battery energy storage to facilitate higher renewable penetration in an isolated power system. *IEEE Access*, 10, 2382–2398. https://doi.org/10.1109/access.2021.3139159

[33]   Ameur, A., Berrada, A., Loudiyi, K., & Adomatis, R. (2021). Performance and energetic modeling of hybrid PV systems coupled with battery energy storage. *Hybrid Energy System Models*, edited by A. Berrada & R. El Mrabet, 195–238. Elsevier, London. https://doi.org/10.1016/b978-0-12-821403-9.00008-1

[34]   Berrada, A., & Mrabet, R. E. (2020). *Hybrid Energy System Models*. Academic Press, Netherlands.

[35]   Xu, C., Behrens, P., Gasper, P., Smith, K., Hu, M., Tukker, A., & Steubing, B. (2023). Electric vehicle batteries alone could satisfy short-term grid storage demand by as early as 2030. *Nature Communications*, 14(1), 119. https://doi.org/10.1038/s41467-022-35393-0

[36]   Pedrero, R. A., De Lestrade, V. V., Specht, J., & Del Granado, P. C. (2023). Value and effects of adopting residential flexibility in the European power system. *2023 IEEE Power & Energy Society General Meeting (PESGM)*. https://doi.org/10.1109/pesgm52003.2023.10252872

[37]   Klingler, A.-L., Schreiber, S., & Louwen, A. (2019). Stationary batteries in the EU countries, Norway and Switzerland: Market shares and system benefits in a Decentralized World. *2019 16th International Conference on the European Energy Market (EEM)*. https://doi.org/10.1109/eem.2019.8916537

[38]   Berrada, A., Loudiyi, K., & El Mrabet, R. (2021). Introduction to hybrid energy systems. *Hybrid Energy System Models*, edited by A. Berrada & R. El Mrabet, 1–43. Elsevier, London. https://doi.org/10.1016/b978-0-12-821403-9.00001-9

[39]   Gholami, M., Sanjari, M., & Berrada, A. (2023). Game theoretical approach for critical sizing of energy storage systems for residential prosumers. *Journal of Energy Storage*, 64, 107166. https://doi.org/10.1016/j.est.2023.107166

[40]   Chatzivasileiadi, A., Ampatzi, E., & Knight, I. P. (2022). Electrical energy storage sizing and space requirements for sub-daily autonomy in residential buildings. *Energies*, 15(3), 1145. https://doi.org/10.3390/en15031145

[41]   Zyglakis, et al. (2021). Energy storage systems in residential applications for optimised economic operation: Design and experimental validation. *2021 IEEE PES Innovative Smart Grid Technologies – Asia (ISGT Asia)*. https://doi.org/10.1109/isgtasia49270.2021.9715588

[42]   Börner, M. F., Frieges, M. H., Späth, B., Spütz, K., Heimes, H. H., Sauer, D. U., & Li, W. (2022). Challenges of second-life concepts for retired electric vehicle

batteries. *Cell Reports Physical Science*, 3(10), 101095. https://doi.org/10.1016/j.xcrp.2022.101095

[43]   Wehner, N., & Daim, T. (2019). Behind-the-meter energy storage implementation. *R&D Management in the Knowledge Era. Innovation, Technology, and Knowledge Management*, edited by T. Daim et al., 71–94. Springer, Cham. https://doi.org/10.1007/978-3-030-15409-7_3

[44]   Singh, B., Tripathi, M., Maithil, S., & Gupta, V. (2023). A review on the integration of electric vehicles into the power grid and its impact on the energy infrastructure in India. *2023 IEEE Renewable Energy and Sustainable E-Mobility Conference (RESEM)*. https://doi.org/10.1109/resem57584.2023.10236371

2 Battery Technologies Comparison
Residential Systems

Cyrus Wekesa

2.1 INTRODUCTION

Energy is fundamental to socio-economic development in every society, with applications in residential, commercial, industrial, transportation, and agricultural sectors. As the world population continues to increase, so does the demand for the energy required to meet everyday needs of the people. This trend has led to increased concern about energy sustainability, security, and affordability. Sustainability aims at decarbonizing energy and combating climate change, while energy security ensures a reliable and uninterrupted supply of energy to meet the energy demand. Energy affordability entails providing energy services at minimal cost to consumers. Solar photovoltaic (PV) has become an important alternative energy source for residential establishments, but the intermittent nature of solar energy requires it to be integrated with energy storage mechanisms such as batteries to guarantee stability in electricity supply.

Residential establishments in modern societies have diverse items of equipment that are powered by electricity to support the needs of modern-day living: major appliances such as water heaters, microwave ovens, refrigerators, clothes dryers, and air-conditioners; small appliances such as electric kettles, blenders, toasters, rice cookers, and meat grinders; and consumer electronics such as video game consoles, computers, digital cameras, clocks, radio receivers, and television sets which are used for recreation, communication, and entertainment. These items of equipment are critical to modern-day life, and therefore a reliable supply of electricity is essential.

Apart from the trend towards the adoption of energy-efficient equipment for residential applications, there is an increase in energy supply technologies including renewable ones in both grid-connected and off-grid residences. Solar PV has become an important alternative energy source for residential establishments, but the intermittent nature of solar energy requires it to be integrated with energy storage mechanisms such as batteries to guarantee stability in electricity supply.

An ideal battery should be cheap, have an unlimited shelf life, operate across a broad range of power levels and temperatures, have almost unlimited energy, and be completely safe. This chapter documents a comprehensive comparison of various battery technologies for residential energy systems. The technologies are compared

16 DOI: 10.1201/9781003441236-2

based on various criteria: technology maturity, discharge duration, reaction time, power range, power density, round-trip efficiency, lifetime, and estimated cost.

2.2 OVERVIEW OF ENERGY STORAGE SYSTEMS

The reliability of energy is critical in any society since energy is needed to meet everyday needs [1–3]. Energy storage is one of the mechanisms used to ensure reliability in energy. Energy storage devices can be classified as electrical, mechanical, electrochemical, chemical, and thermal, as illustrated in Figure 2.1.

The storage technologies are compared based on technical and economic factors, and they truly have different technical and cost characteristics. For instance, storage technologies like PHS and CAES are site-specific, while battery energy storage systems (BESS) are flexible in terms of their location. BESS also has a promising cost characteristic going into the future compared to other energy storage technologies. The increase in mass production of secondary batteries, the decrease in their cost, and the recent growth in electric vehicles and mobile information technology (IT) devices have led to an increase in demand for energy storage systems that use such batteries [2, 3]. Using electrochemical methods, secondary batteries such as lithium batteries, lead storage batteries, and sodium-sulfur batteries transform electric energy into chemical energy and vice versa. Other energy storage technologies include double-layer capacitors (electrical) and fuel cells (chemical) [4, 5].

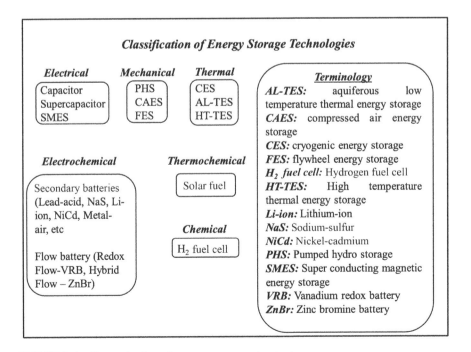

FIGURE 2.1   Categorization of energy storage technologies.

Batteries can be used at grid-level electrical energy storage because of their favorable characteristics such as flexible installation, modularization, quick reaction, and short construction cycles [6–8]. They also meet the requirements for grid energy storage applications such as capacity, energy efficiency (EE), longevity, power density, and energy density [9, 10]. Grid functions such as frequency management, peak shaving, load leveling, and extensive integration of renewable energies must be considered when analyzing batteries to be used in grid-level energy storage systems.

Traditional applications of battery storage have been in uninterruptible power supplies (UPS), renewable energy systems, and battery-powered devices like cell phones, computers (laptops), cameras, drills, and screwdrivers. Due to new applications of battery storage technologies such as facilitating large-scale integration of renewable energy into the power grid, there is a need for further research on battery energy storage technologies to improve round-trip efficiency and other technical and cost factors [11, 12]. This will enable battery storage technology to be useful across the entire spectrum, from large power systems to small residential systems. In the event of planned or unforeseen energy load shedding or power outages, battery technology is employed as a backup source at the home level [12–16]. Specifically, in Karachi, Pakistan, the total annual energy consumption in the home sector was 3989 GWh in 2013, with 13.1 percent of the total energy used to charge UPSs [12].

2.3   CLASSIFICATION AND ROLE OF BATTERY STORAGE SYSTEMS

Batteries are typically divided into primary batteries and secondary batteries, with the former corresponding to non-rechargeable batteries and the latter corresponding to rechargeable batteries [17–19]. Primary batteries are utilized in manufacturing environments and various low- to medium-power electronic devices like watches, radios, and calculators, while secondary batteries are commonly used as BESS [20] in a wide variety of applications:

- Stationary applications: off-grid energy storage, UPSs, emergency power, load leveling, peak shaving, load shifting, and dispatchability of non-programmable renewable energy plants;
- Automotive sector in electric and hybrid vehicles; and
- Portable devices such as laptops, mobile phones, and cordless devices.

It is worth noting the important role that batteries play in off-grid energy systems. An estimated 1 billion people around the world do not have access to power grids. To meet their energy needs, they frequently rely on expensive and polluting diesel generators. Stand-alone systems based on renewable energy sources (RES) are an important development to address the challenge of cost and pollution from these generators. Due to the variability and intermittent nature of RES, BESS is an essential component of the stand-alone systems to store and provide energy according to the owners' needs and to ensure continuous power supply.

The use of BESS boosts the amount of renewable energy that can be implemented in off-grid systems up to 100 percent, ensuring the delivery of clean and sustainable energy [21–23]. According to Tianmei Chen *et al.* [4], economic growth and

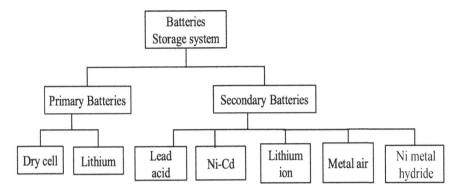

FIGURE 2.2 Battery classification based on usage.

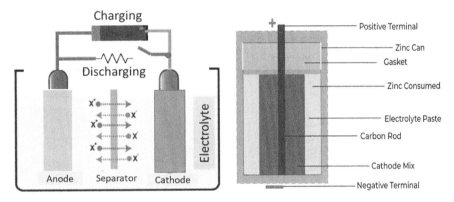

FIGURE 2.3 A schematic illustration of (a) secondary cell, (b) primary cell.

social prosperity are currently more dependent on electrical energy compared to the past, which leads to strong demand for a grid-level energy storage system. Figure 2.2 shows battery classification based on usage cycle (chargeable and non-rechargeable). Figure 2.3 provides a schematic view of both primary and secondary batteries. Figures 2.4 and 2.5 show an application of battery storage technology in on-grid and off-grid scenarios, respectively.

Primary batteries are used where power requirement is low (memory back-ups, remote controllers, and flashlight) and in the case where the battery is used infrequently over a long period. Secondary batteries are used in cases of high discharge rates such as portable tools, electric bikes, automobiles, and utility systems.

2.4 BATTERY TECHNOLOGY AND CHARACTERISTICS

Electrochemical cells and batteries can be categorized as primary (non-rechargeable) and secondary (rechargeable). The variants of each are generated from this basic categorization [19].

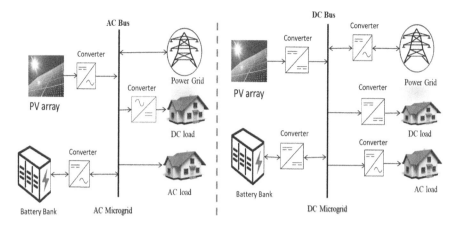

FIGURE 2.4   Secondary battery integrated with on-grid PV system.

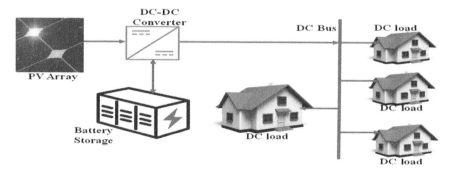

FIGURE 2.5   Off-grid PV-battery integration at residential systems.

2.4.1 PRIMARY BATTERIES

The general features as well as applications of the various primary battery types are presented in what follows.

Zinc-Carbon Batteries

The zinc-carbon dry cell battery has existed for over 100 years and has been the most widely used of all the dry cell batteries because of its low cost, relatively good performance, and ready availability. Cells and batteries of many sizes and characteristics have been manufactured to meet the requirements of a wide variety of applications. Significant improvements in capacity and shelf life were made in this battery system in the period between 1945 and 1965 through the use of new materials (such as beneficiated manganese dioxide and zinc chloride electrolyte) and cell designs. The low cost of the zinc-carbon battery is a major attraction, but it has lost a considerable market share, except in developing countries, because of the newer primary batteries with superior performance characteristics [19, 28].

Zinc/Alkaline/Manganese Dioxide Batteries

The Zn/alkaline/MnO_2 battery has had a firm footing on the primary battery market for a while now and has replaced other batteries because of its improved performance at higher current demands and low temperatures, with a longer lifespan. For applications requiring high-rate or low-temperature capability, it is cost-effective even though more expensive per unit than the lead-acid battery [19, 28, 29].

Zinc/Mercuric Oxide Batteries

Another zinc anode primary battery system is the zinc/mercuric oxide battery. It has a long lifespan and high internal energy density. It is an important power source for electronic watches, calculators, hearing aids, photographic equipment, and other devices that need a dependable, long-lasting micro power source. However, it has now been largely replaced by other battery systems, such as zinc/air and lithium batteries, which offer superior performance in a wide range of applications [19, 28, 29].

Zinc/Silver Oxide Batteries

The primary zinc/silver oxide battery is similar in design to the small zinc/mercuric oxide button cell, but it has a higher energy density (on a weight basis) and performs better at low temperatures. These characteristics make this battery system desirable for use in hearing aids, photographic applications, and electronic watches. However, because of its high cost and the development of other battery systems, the use of the zinc/silver oxide battery system, as a primary battery, has been limited mainly to small button battery applications where the higher cost is justified [19, 29].

Zinc/Air Batteries

Despite being known for its great energy density, the zinc/air battery technology had only been applied to big, low-power batteries for signaling and directional aid applications. The high-rate capabilities of the system were enhanced with the invention of better air electrodes, and compact button-type batteries which are now commonly employed in hearing aids, electronics, and related applications. These batteries have a very high energy density because active cathode material is not required. Performance constraints such as sensitivity to severe temperatures and humidity, as well as limited activate shelf life and low power density, have contributed to this system not being widely used, and the development of larger batteries has been delayed [2, 19, 28, 29].

Lithium Batteries

Lithium batteries are gradually replacing traditional battery systems because they offer the best energy density, can operate over a wide variety of temperatures, and have a long lifespan [21], [23]. In applications such as watches, calculators, memory circuits, photography equipment, and communication devices where high energy density and long shelf life are crucial, lithium batteries have replaced the standard primary batteries. These batteries, which come in sizes up to roughly 35 Ah, are utilized in lighting goods, industrial and military applications, and other equipment where compactness, lightweight, and functioning over a wide temperature range are crucial [19, 29].

Figure 2.6 shows in picture form the different types of primary batteries [30].

FIGURE 2.6 Examples of primary battery types and sizes.

2.4.2 Secondary Batteries: Basic Types and Characteristics

Secondary batteries, described next, consist of a set of cells with reversible cell processes, with the original chemical conditions inside the cell being restored through charging.

Lead-Acid (LA) Batteries

Lead-acid batteries have been the most popular type of battery used worldwide. The normal service life is 3–6 years, with efficiency in the range of 80 to 90 percent [31–34]. The drawbacks include reduced energy density and the usage of lead, a hazardous substance that is sometimes outlawed or regulated. The benefits of lead-acid batteries include a favorable cost–performance ratio, straightforward recycling, and an easy charging process. The improvement of lead-acid batteries' efficiency for micro-hybrid electric cars is the current area of focus. The battery schematic is shown in Figure 2.7.

Nickel-Cadmium and Nickel-Metal Hydride Batteries

Nickel-cadmium (NiCd) batteries have been in use for more than a century, while nickel-metal hydride (NiMH) batteries first went on sale more than two decades ago. Except for a tenfold lower maximum nominal capacity, NiMH batteries have all the benefits of NiCd batteries, including greater power density, somewhat better energy density, and a higher number of cycles. They outperform lithium-ion batteries in terms of strength and security. The use of nickel- cadmium is now restricted because of cadmium's toxicity. Currently, the price of NiMH batteries is comparable to that of Li-ion battery packs [23].

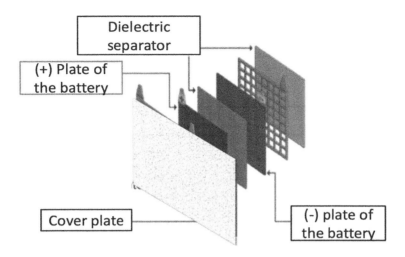

FIGURE 2.7   Schematics of lead-acid battery cells.

Lithium-Ion (Li-Ion) Batteries

Lithium-ion batteries currently dominate the portable and mobile energy storage market. One lithium-ion cell can replace three NiCd and NiMH batteries at a cell voltage of just 1.2 V [23]. The main barrier is the high price of the special packaging and integrated overload protection circuits. Lithium-ion battery technology has a significant safety issue since the majority of metal oxide electrodes are thermally unstable and can melt in hot conditions. A monitoring system built into lithium-ion batteries guards against overcharging and over-discharging to reduce this risk. A voltage regulation circuit is frequently supplied to track and prevent voltage variations in individual cells. There are numerous opportunities for progress in the development of lithium-ion battery cathodes [35, 36]. A schematic diagram of the lithium-ion battery is shown in Figure 2.8.

Sodium-Sulfur (NaS) Batteries

In sodium-sulfur batteries, the active components—molten sodium at the cathode and molten sulfur at the anode—are separated by a solid beta-alumina ceramic electrolyte, and the battery has a typical life cycle of about 4500 cycles and a discharge time of 6.0 to 7.2 hours. They are quick to react, with a round-trip AC efficiency of about 75 percent [33]. The absence of a heat source is a considerable disadvantage, but the heat generated in the battery can be controlled with frequent use of its reaction heat with the right insulating size [33, 35, 36]. Applications requiring high-frequency cycling are ideal for sodium-sulfur batteries. The construction of a typical NaS battery module is illustrated in Figure 2.9.

2.5   SUMMARY OF RESIDENTIAL APPLICATIONS OF BATTERIES

Diverse products and systems use batteries; products refer to individual devices powered by batteries, whether portable or not, while systems refer to large installations

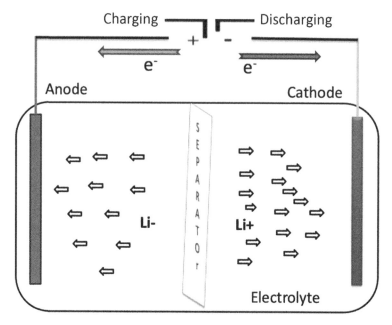

FIGURE 2.8   Schematic block of Li-ion battery.

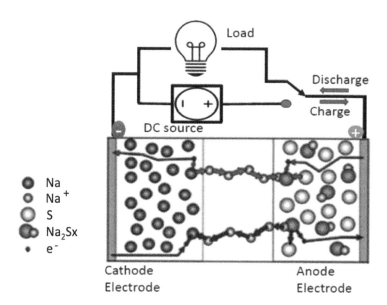

FIGURE 2.9   Schematic diagram of sodium-sulfur battery.

such as energy storage plants to back up power from the electricity grid. Battery applications can be classified as portable (mostly consumer applications), industrial (telecommunications and weather forecasting), and traction/automotive (electric and hybrid electric cars). Tables 2.1 and 2.2 capture the applications of primary and secondary batteries, respectively, and their main features. From the tables, it is clear that the requirements for battery power vary and so do the environmental and electrical conditions under which batteries have to operate. Thus, there is no ideal battery that can meet all the diverse requirements [1, 11, 19, 26, 37].

From the diverse applications captured, it is clear that batteries play a significant role in our daily lives as a source of electrical energy. Primary batteries are typically used once and discarded, meaning that applications that require a long lifetime should consider the use of secondary batteries. While primary batteries cost less per battery and have a higher energy density, they are an expensive option over time.

TABLE 2.1
Primary batteries: Applications and characteristics

S/N	Battery type	Application areas	Characteristics
1	Zinc-carbon	Flashlights, portable radios, toys, novelties, instruments	Common; low-cost primary battery; available in a variety of sizes
2	Mercury (Zn/HgO)	Hearing aids, medical devices (pacemakers), photography, detectors, and military equipment but in limited use due to the environmental hazard of mercury	Highest capacity (by volume) of conventional types; flat discharge; good shelf life
3	Alkaline (Zn/alkaline/MnO_2)	Most popular primary battery; used in a variety of portable battery-operated equipment	Most popular general-purpose premium battery; good low-temperature and high-rate performance; moderate cost
4	Silver/zinc (Zn/Ag_2O)	Hearing aids, photography, electric watches, missiles, underwater and space applications (larger sizes)	Highest capacity (by weight) of conventional types; flat discharge; good shelf life; costly
5	Zinc/air (Zn/O_2)	Special applications, hearing aids, pagers, medical devices, portable electronics	Highest energy density; low cost; not independent of environmental conditions
6	Lithium/solid cathode	Replacement for conventional button and cylindrical cell applications	High energy density; good rate capability and low-temperature performance; long shelf life; competitive cost
7	Lithium/solid electrolyte	Medical electronics, memory circuits, fusing	Extremely long shelf life; low-power battery

TABLE 2.2
Secondary batteries: Applications and characteristics

S/N	Battery type	Applications area	Characteristics
1	Lead-acid (Stationary)	Emergency power, utilities, telephone, UPS, load leveling, energy storage, emergency lighting	Designed for standby float service; long life; VRLA designs
2	Lead-acid (Portable)	Portable tools, small appliances and devices, TV, and portable electronic equipment	Sealed; maintenance-free; low cost; good float capability; moderate cycle life
3	Nickel-cadmium (Portable)	Consumer electronics, portable tools, pagers, appliances, photographic equipment, standby power, memory backup	Sealed; maintenance-free; good high-rate low-temperature performance; excellent cycle life
4	Nickel-metal hydride	Consumer electronics and other portable applications; electric and hybrid electric vehicles	Sealed; maintenance-free; higher capacity than nickel-cadmium batteries
5	Lithium-ion	Portable and consumer electronic equipment, electric vehicles, and space applications	High specific energy and energy density; long cycle life
6	Sodium-sulfur	Electric grid, or stand-alone renewable power applications	High efficiency and long life cycle; high heat generation

2.6   COMPARISON OF BATTERY TECHNOLOGIES

As previously mentioned, batteries can be compared based on several technical and cost factors. Table 2.3 provides a comparison of several primary batteries, while Table 2.4 provides a comparison of secondary battery technologies [19, 29].

The lead-acid battery is a mature technology with very low cost, as seen in Table 2.4, but has several drawbacks, including low energy density, shorter lifetime, and the usage of lead, which is considered a hazardous substance and is sometimes outlawed or regulated. On the other hand, the sodium-sulfur battery is in the initial commercialization development stage. The Li-ion battery has a striking advantage in terms of gravimetric and volumetric energy densities. Gravimetric energy density expresses the amount of energy of a battery per unit weight (Wh/kg), whereas volumetric energy density expresses the energy per unit volume (Wh per liter, Wh/L). The high gravimetric/volumetric energy density of the Li-ion battery means that it carries a very large amount of energy in a small package, allowing products powered by Li-ion batteries to be smaller and lighter. This is important in high-end and costlier cellular phones and laptop computers due to the high cost of the battery [2, 3, 11, 19, 24–27, 36, 38–40].

Relative to lead-acid batteries, Li-ion batteries have higher energy efficiency, higher energy density, higher life cycle, and lower maintenance. On the other hand,

TABLE 2.3
Primary battery energy storage technology comparison

Battery Type	Voltage in a single cell [V]	Specific energy (Wh/kg)	Energy density (Wh/L)	Power density (W/L)	Lifetime rank
Zinc-carbon	1.5	85	165	4	6
Mercury (Zn/HgO)	1.35	100	470	2	3
Alkaline (Zn/alkaline/MnO$_2$)	1.5	145	400	2	5
Silver/zinc (Zn/Ag$_2$O)	1.6	135	525	2	4
Zinc/air (Zn/O$_2$)	1.5	370	1300	3	—
Lithium/solid cathode	3.6	590	1100	1	1
Lithium/solid electrolyte	1.5	260	500	5	2

TABLE 2.4
Secondary battery energy storage technology comparison

	Battery type				
Key comparison parameter	Lead-acid	Nickel-cadmium (Portable)	Nickel-metal hydride	Lithium-ion	Sodium-sulfur
Single cell voltage [V]	2	1.2	1.2	4/1	2
Specific energy (Wh/kg)	35	35	75	150	170
Energy density (Wh/L)	50–80	40–60	40–60	200–400	150–300
Power density (W/L)	90–700	120–850	550–1000	1300–10,000	120–160
Lifetime (years)	3–6	10–20	10–20	10–15	10–15
Efficiency (%)	80–90	60–80	80–90	85–95	75–90
Cost ($/KWh)	200–400	800–1500	400–700	900–6200	445–555
Discharge time	1 min–8 hrs	Secs–hours	Sec–hours	1 min–8 hrs	Sec–hours

owing to the affordable costs of lead-acid batteries for large installed capacities, they are the dominant technology for off-grid solar photovoltaic applications. The Li-ion battery also has a high nominal voltage relative to the others, as shown in Table 2.4, meaning that a designer will need several other cells connected in series to equal the voltage of a single Li-ion cell.

2.7 CURRENT AND FUTURE OUTLOOK ON BATTERY TECHNOLOGIES

Resulting from their long life cycle, higher energy density, low maintenance requirements, and high electrochemical potential, lithium-ion batteries are the preferred choice for use in smartphones and consumer electronic products. Their versatility allows them to be used in very small packs as well as in large-scale installations. The upsides of these batteries have seen them displace nickel-cadmium (NiCd) batteries in key applications such as portable electronic devices (smartphones and laptops) found in residential establishments. Figure 2.10, modified from [41], prominently illustrates one such upside: high gravimetric (specific) and volumetric energy densities.

With the increasing popularity of consumer electronics globally, it is expected that the battery market size will continue to grow. In fact, between years 2020 and 2030, this market is expected to grow at a combined annual growth rate (CAGR) of 14.1 percent in the USA [42]. More significantly, lithium-ion batteries will take up applications currently served by lead-acid batteries (in electric vehicles, storage, etc.). Furthermore, high demand for portable electronics—smartphones, tablets, and wearable devices—will drive further growth of lithium-ion. Finally, the shift from fossil fuels to renewable energy, including the growth of the e-mobility industry, will significantly grow the market for batteries, where lithium-ion batteries are expected to dominate.

Globally, the annual battery demand continues to be fueled by motor vehicles, electrical and electronic equipment, and motive power equipment. As global income levels rise due to economic development, the market is expected to grow further [43]. The growth of electric vehicles, the transition from fossil fuels to RES, and the increase in mobile IT devices will all contribute to increased demand for batteries, a

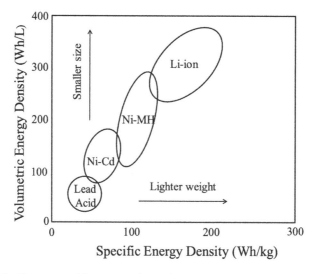

FIGURE 2.10   Battery-specific versus volumetric energy densities.

trend that will make batteries more significant in our lives than ever before. Several battery technology performance characteristics such as energy density, power density, life span, cost, safety, and charging time need to be improved [44], calling for further research into battery technologies.

Innovations are needed in the following aspects [42–44]: development of higher-capacity batteries for electric vehicles and advanced battery management systems (safety and efficiency improvement); further research and development of solid-state lithium-ion batteries to improve safety, energy density, and charging times; and recycling of lithium-ion batteries to reduce the environmental impact of batteries and to provide a sustainable source of raw materials for new batteries.

2.8　CONCLUSION

A wide variety of consumer, industrial, and military applications require power and batteries as an important source of energy. Advancing electronics technology and the continuing development of portable devices—on top of electric vehicles and utility power load leveling applications—means that batteries are expected to continue playing an increasingly significant role in people's lives. Battery selection is an important consideration and factors like access to the product, temperature sensitivity, initial cost, maintenance, weight, and volume require careful consideration because there is no single battery that can be a solution for every application or requirement.

Where the power requirement is low—or in applications in which a battery will be used infrequently for a long period—the primary battery is preferred. As power requirements increase and the size of the battery becomes larger, the secondary battery is preferred. In between these two cases, the choice could depend on user preference: balancing between the convenience of the primary battery and freedom from the charger and the inconvenience of the rechargeable battery and potentially lower operating costs.

REFERENCES

[1]　IEA, "World Energy Outlook 2019," *World Energy Outlook 2019*, p. 1, 2019, [Online]. Available: www.iea.org/weo%0Ahttps://www.iea.org/reports/world-energy-outlook-2019%0Ahttps://www.iea.org/reports/world-energy-outlook-2019%0Ahttps://webstore.iea.org/download/summary/2467?fileName=Japanese-Summary-WEO2019.pdf

[2]　P. Blanc, M. Zafar, J. Levy, and P. Gupta, "Five Steps to Energy Storage. Innovation Insights Brief 2020," *World Energy Council*, p. 62, 2020, [Online]. Available: www.worldenergy.org

[3]　M. R. Chakraborty, S. Dawn, P. K. Saha, J. B. Basu, and T. S. Ustun, "A comparative review on energy storage systems and their application in deregulated systems," *Batteries*, vol. 8, no. 9, 2022, doi: 10.3390/batteries8090124

[4]　T. Chen *et al.*, "Applications of lithium-ion batteries in grid-scale energy storage systems," *Trans. Tianjin Univ.*, vol. 26, no. 3, pp. 208–217, 2020, doi: 10.1007/s12209-020-00236-w

[5]　M. M. Symeonidou, E. Giama, and A. M. Papadopoulos, "Life cycle assessment for supporting dimensioning battery storage systems in micro-grids for residential applications," *Energies*, vol. 14, no. 19, 2021, doi: 10.3390/en14196189

[6]   K. Antoniadou-Plytaria, D. Steen, L. A. Tuan, O. Carlson, and M. A. Fotouhi Ghazvini, "Market-based energy management model of a building microgrid considering battery degradation," *IEEE Trans. Smart Grid*, vol. 12, no. 2, pp. 1794–1804, 2021, doi: 10.1109/TSG.2020.3037120

[7]   C. Jamroen, N. Yonsiri, T. Odthon, N. Wisitthiwong, and S. Janreung, "A standalone photovoltaic/battery energy-powered water quality monitoring system based on narrowband internet of things for aquaculture: Design and implementation," *Smart Agric. Technol.*, vol. 3, no. May 2022, p. 100072, 2023, doi: 10.1016/j.atech.2022.100072

[8]   A. Kapoor and A. Sharma, "Optimal charge/discharge scheduling of battery storage interconnected with residential PV system," *IEEE Syst. J.*, vol. 14, no. 3, pp. 3825–3835, 2020, doi: 10.1109/JSYST.2019.2959205

[9]   M. D. Anderson and D. S. Carr, "Battery energy storage technologies," *Proc. IEEE*, vol. 81, no. 3, pp. 475–479, 1993, doi: 10.1109/5.241482

[10]   D. W. Sobieski and M. P. Bhavaraju, "An economic assessment of battery storage in electric utility systems," *IEEE Trans. Power Appar. Syst.*, vol. PAS-104, no. 12, pp. 3453–3459, 1985, doi: 10.1109/TPAS.1985.318895

[11]   A. J. and M. Shahidehpour, "Battery storage systems in electric power systems," *2006 IEEE.*, vol. 1, pp. 1279–1286, 2006.

[12]   J. A. Qureshi, T. T. Lie, K. Gunawardane, N. Kularatna, and W. A. Qureshi, "AC source vs DC source: Charging efficiency in battery storage systems for residential houses," *2017 IEEE Innov. Smart Grid Technol. – Asia Smart Grid Smart Community, ISGT-Asia 2017*, pp. 1–6, 2018, doi: 10.1109/ISGT-Asia.2017.8378378

[13]   A. Chatzivasileiadi, E. Ampatzi, and I. P. Knight, "The choice and architectural requirements of battery storage technologies in residential buildings," *Second International Conference for Sustainable Design of the Built Environment (SDBE 2018), London*, 2018, https://orca.cardiff.ac.uk/id/eprint/114033/1/102_v2.pdf

[14]   A. M. Abdelshafy, H. Hassan, A. M. Mohamed, G. El-Saady, and S. Ookawara, "Optimal grid connected hybrid energy system for Egyptian residential area," *Proceeding – ICSEEA 2017 Int. Conf. Sustain. Energy Eng. Appl. Continuous Improv. Sustain. Energy Eco-Mobility*, vol. 2, pp. 52–60, 2017, doi: 10.1109/ICSEEA.2017.8267687

[15]   S. D. Percy, M. Aldeen, C. N. Rowe, and A. Berry, "A comparison between capacity, cost, and degradation in Australian residential battery systems," *IEEE Innovative Smart Grid Technologies Asia (ISGT-Asia), Melbourne, 2016*, pp. 202–207, 2016, doi: 10.1109/ISGT-Asia.2016.7796386

[16]   M. Alramlawi and P. Li, "Design optimization of a residential PV-battery microgrid with a detailed battery lifetime estimation model," *IEEE Trans. Ind. Appl.*, vol. 56, no. 2, pp. 2020–2030, 2020, doi: 10.1109/TIA.2020.2965894

[17]   S. Corigliano and A. Cortazzi, "Battery energy storage systems modeling for robust design of microgrids in developing countries," 2017, [Online]. Available: www.polit esi.polimi.it/handle/10589/137750

[18]   S. Bhosale, "Week 1 Understanding Different Battery Chemistry," *SKILL LYNC.* https://skill-lync.com/student-projects/week-1-understanding-different-battery-chemistry-583 (accessed Jun. 06, 2023).

[19]   D. Linden and T. B. Reddy, *Handbook of batteries*, 3rd ed. New York, Chicago: McGraw-Hill New, 2002. doi: 10.1002/9780470933886.ch1

[20]   Martin Winter and Ralph J. Brodd, "What are batteries, fuel cells, and supercapacitors?," *ACS*, vol. 10, no. 104, pp. 4245–4270, 2004.

[21]   J. Warner., *The Handbook of Lithium-Ion battery pack design*. Amsterdam: Elsevier, 2015.

[22] X. Luo, J. Wang, M. Dooner, and J. Clarke, "Overview of current development in electrical energy storage technologies and the application potential in power system operation," *Appl. Energy*, vol. 137, pp. 511–536, 2015, doi: 10.1016/j.apenergy.2014.09.081

[23] K. Hardik *et al.*, "Comparison of lead-acid and lithium ion batteries for stationary storage in off-grid energy systems," *IEEE PES Innov. Smart Grid Technol. Conf. Eur.*, vol. 2, no. 11, pp. 125–259, 2020.

[24] S. Bandyopadhyay, Z. Qin, L. Ramirez-Elizondo, and P. Bauer, "Comparison of battery technologies for DC microgrids with integrated PV," 2019 IEEE Third International Conference on DC Microgrids (ICDCM), *Matsue, Japan, 2019*, pp. 1–9, 2020, doi: 10.1109/icdcm45535.2019.9232746

[25] A. Z. Gabr, A. A. Helal, and N. H. Abbasy, "Multiobjective optimization of photo voltaic battery system sizing for grid-connected residential prosumers under time-of-use tariff structures," *IEEE Access*, vol. 9, pp. 74977–74988, 2021, doi: 10.1109/ACCESS.2021.3081395

[26] A. Clean and E. Forum, "Energy storage system technology and demonstration," *Asia Clean Energy Forum 2018*, p. 18971, Jun. 2018, [Online]. Available: www.adb.org

[27] E. Martinez-Laserna *et al.*, "Technical viability of battery second life: A study from the ageing perspective," *IEEE Trans. Ind. Appl.*, vol. 54, no. 3, pp. 2703–2713, 2018, doi: 10.1109/TIA.2018.2801262

[28] B. Viswanathan and B. Viswanathan, Chapter 12 – Batteries. *Energy sources*, ed. B. Viswanathan, Elsevier, Amsterdam, 2017. doi: 10.1016/B978-0-444-56353-8.00012-5

[29] R. Higgins and K. Kruger, "High energy non-rechargeable batteries and their applications," *SAE Technical Paper* 901052, 1990, doi: 10.4271/901052

[30] B. Learning, "Difference between Primary Cell and Secondary Cell," *BYJU'S Learning*, 2019. https://byjus.com/chemistry/difference-between-primary-cell-and-secondary-cell/ (accessed Jul. 06, 2023).

[31] C. H. Dustmann, "Advances in ZEBRA batteries," *J. Power Sources*, vol. 127, no. 1–2, pp. 85–92, 2004, doi: 10.1016/j.jpowsour.2003.09.039

[32] H. Keshan, J. Thornburg, and T. S. Ustun, "Comparison of lead-acid and lithium-ion batteries for stationary storage in off-grid energy systems," *4th IET Clean Energy and Technology Conference (CEAT 2016)*, IET Library Digital, 2016, pp. 14–15.

[33] E. M. G. Rodrigues *et al.*, "Modelling and sizing of NaS (sodium sulfur) battery energy storage system for extending wind power performance in Crete Island," *Energy*, vol. 90, pp. 1606–1617, 2015, doi: 10.1016/j.energy.2015.06.116

[34] U. G. K. M. and W. Shen, "A review of battery energy storage systems for residential DC microgrids and their economical comparisons," *2018 Jt. Int. Conf. Energy, Ecol. Environ. (ICE 2018)*, vol. 10, no. 127, 27856, 2019.

[35] M. Götz *et al.*, "Renewable power-to-gas: A technological and economic review," *Renew. Energy*, vol. 85, pp. 1371–1390, 2016, doi: 10.1016/j.renene.2015.07.066

[36] IEC, "Electrical energy storage: Technology overview and applications," *Electron. Energy Storage White Pap.*, vol. 39, no. July, pp. 11–12, 2009.

[37] N. Saxena, I. Hussain, B. Singh, and A. L. Vyas, "Implementation of a grid-integrated PV-battery system for residential and electrical vehicle applications," *IEEE Trans. Ind. Electron.*, vol. 65, no. 8, pp. 6592–6601, 2018, doi: 10.1109/TIE.2017.2739712

[38] O. Ovwigho and A. Ianga, "Comparative analysis of battery storage technologies for residential photovoltaic solar energy installations," *Int. J. Res. Find. Eng. Sci. Technol.*, vol. 4, no. 2, pp. 26–40, 2022, doi: 10.48028/iiprds/ijrfest.v4.i2.03

[39]   M.-C. L. Chao Zhang, Yi-Li Wei, Peng-Fei Cao, "Energy storage system: Current studies on batteries and power condition system," *Renewab Sustainable Ener Rev.*, vol. 82, pp. 3091–3106, 2018.

[40]   BRE & RECC, "Batteries and solar power: Guidance for domestic and small commercial Consumers." BRE National Solar Centre, 2016, https://files.bregroup.com/bre-co-uk-file-library-copy/filelibrary/nsc/Documents%20Library/NSC%20Publications/88166-BRE_Solar-Consumer-Guide-A4-12pp-JAN16.pdf (accessed Jun. 15, 2024).

[41]   "Global Consumer Battery Market Size Analysis Report, 2030." www.grandviewr esearch.com/industry-analysis/consumer-battery-market-report (accessed Jun. 15, 2023).

[42]   "Clean Energy Institute (CEI) at the University of Washington," 2020. www.cei. washington.edu/education/science-of-solar/battery-technology/ (accessed Jun. 15, 2023).

[43]   "Global Batteries Industry Reports, Global." www.freedoniagroup.com/industry-study/global-batteries-4412.htm (accessed Jun. 15, 2023).

[44]   "The Future of Battery Technology," 2020. www.spglobal.com/esg/s1/topic/the-fut ure-of-battery-technology.html (accessed Jun. 15, 2023).

3 A Review of Lithium-Ion Battery Models

Wei Liu, Jiashen Teh, and Jian Shi

3.1 INTRODUCTION

With the gradual depletion of fossil fuels, countries have increasingly focused on energy crisis and environmental issues [1], which has led to rapid development in the fields of renewable energy, electric vehicles, smart grids, and energy storage technologies [2–4]. Among the above-mentioned application areas, the battery energy storage system (BESS) is one of the key technologies [5, 6]. LIBs, with their high energy density, long cycle life, and no memory effect, have been widely used in BESS [7]. However, there are some issues related to the safety, durability, uniformity, and cost of LIBs, thus requiring them to operate within safe and reliable ranges of current, voltage, temperature, etc. To achieve this goal, BMS are utilized to manage and estimate the state of the batteries [8]. The key states of the battery include state of charge (SOC), state of health (SOH), state of power (SOP), state of energy (SOE), and state of temperature (SOT) [9, 10]. When estimating the battery states in the BMS, it is necessary to rely on appropriate battery models and related state estimation algorithms. Although the internal physical and chemical reactions of LIBs are complex and there are fewer measurable physical quantities, different states of LIBs have different time scales, as shown in Figure 3.1 [11]. This enables the BMS to model the battery and estimate its state by employing specific testing and modeling methods. Therefore, this chapter provides a review of commonly used battery models, including EMs, ECMs, and TMs.

The remaining sections of this chapter are organized as follows. Section 3.2 provides a review of EMs, which describe the characteristics and reactions of LIBs and belong to battery mechanistic models with complex structures. Section 3.3 presents a review of ECMs, which utilize voltage sources, capacitors, resistors, and other lumped components to describe the variations in battery characteristics. Section 3.4 provides a review of TMs, which are primarily used to describe the temperature variations of batteries and are often combined with EMs or ECMs to capture the thermal-electric coupling characteristics. Finally, Section 3.5 concludes the chapter by summarizing the main findings.

DOI: 10.1201/9781003441236-3

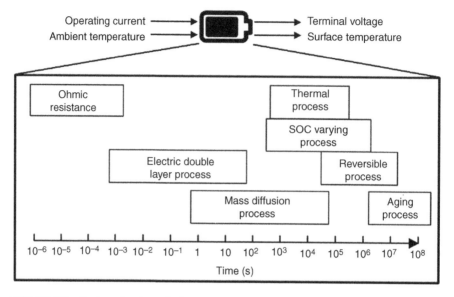

FIGURE 3.1   Systematic view of a battery.

3.2   ELECTROCHEMICAL MODELS

EMs encompass the internal transport processes within a LIB, including lithium-ion transport, electrochemical kinetics, and material properties [12]. As a result, EMs can be utilized for battery electrode concentration estimation, battery state estimation, and charging strategy optimization [13]. Currently, EMs for LIBs can be classified into four main categories: pseudo two-dimensional (P2D) model, single particle (SP) model, enhanced single particle (ESP) model, and physics-based fractional-order model (PBFOM). The purpose of reconstructing these models is to maintain model accuracy while reducing computational complexity, enabling real-time estimation of battery states based on EMs.

3.2.1   Pseudo Two-Dimensional Model

The P2D model is an electrochemical model based on concentrated solution theory, porous electrode theory, and kinetic equations. It has been widely used in battery research and exhibits relatively high model accuracy. The structure of the P2D model is shown in Figure 3.2 [14, 15]. The full-order P2D model consists of porous positive and negative electrodes, a separator, and current collector on both sides. The P2D model is based on three fundamental assumptions [16, 17]: first, it neglects the double-layer effect; second, the electrode material is assumed to be uniformly distributed spherical particles; and third, it considers the kinetics of electrochemical reactions in only one direction along the current collector. The dynamic performance of the battery is described and characterized by a series of partial differential

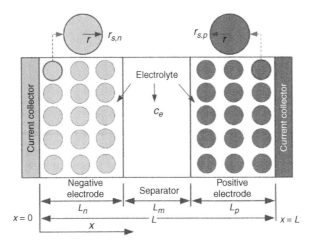

FIGURE 3.2 Schematic diagram of P2D model.

equations (PDEs), including the charge and mass conservation equations for the solid phase, the charge and mass conservation equations for the electrolyte, and the kinetics equations for electrode electrochemical reactions. These equations capture the internal dynamic behavior of the electrode in both the x-direction (electrode thickness) and the r-direction (electrode particle size). Furthermore, the governing equations and boundary conditions of the P2D model are summarized in Table 3.1, while the detailed derivation of the equations and the meanings of various parameters can be found in Table 3.2 and referenced literature [14, 18].

The P2D model is composed of numerous high-order non-linear equations and PDEs, which often require iterative calculations for solving. Despite its high accuracy, the computational process of the model is lengthy and constrained by computational efficiency and variations in parameter identification conditions. To meet the requirements of engineering applications, researchers have been continuously exploring methods to simplify the P2D model.

3.2.2 Single Particle Model

The SP model is a simplified form of the P2D model, where each electrode is treated as a single spherical particle. It assumes a uniform distribution of current throughout the entire electrode and neglects the limitations of electrolyte transport, as shown in Figure 3.3 [19]. This simplification significantly reduces the computational cost of the model while still preserving the internal electrochemical reaction processes. Therefore, the SP model has been widely applied in fields such as battery state estimation [20].

Furthermore, since the SP model assumes a uniform distribution of lithium-ion flux on the porous electrode, the average pore wall flux can be represented. The specific expression for this can be found in Equation (3.1) [21].

TABLE 3.1
The governing equations of the P2D model for LIB

Physical process	Governing equation	Boundary condition	Equation
Lithium-ion diffusion in the solid phase	$\dfrac{\partial c_s}{\partial t} = \dfrac{D_s}{r^2}\dfrac{\partial}{\partial r}\left(r^2\dfrac{\partial c_s}{\partial r}\right)$	$\left.\dfrac{\partial c_s}{\partial r}\right\|_{r=0} = 0,\ D_s\left.\dfrac{\partial c_s}{\partial r}\right\|_{r=r_s} = -\dfrac{j_{li}}{a_s F}$	(3.1)
Lithium-ion diffusion in the electrolyte phase	$\varepsilon_e\dfrac{\partial c_e}{\partial t} = D_e^{eff}\dfrac{\partial^2 c_e}{\partial x^2} + \dfrac{a_s\left(1-t_+^0\right)}{F}j_{li}$	$\left.\dfrac{\partial c_e}{\partial x}\right\|_{x=0} = \left.\dfrac{\partial c_e}{\partial x}\right\|_{x=L} = 0$	(3.2)
Charge conservation in the solid phase	$\sigma^{eff}\dfrac{\partial^2\phi_s}{\partial x^2} = a_s j_{li}$	$-\sigma_-^{eff}\left.\dfrac{\partial\phi_s}{\partial x}\right\|_{x=0} = \sigma_+^{eff}\left.\dfrac{\partial\phi_s}{\partial x}\right\|_{x=L} = \dfrac{I}{A}$	(3.3)
Charge conservation in the electrolyte phase	$k^{eff}\dfrac{\partial^2\phi_e}{\partial x^2} + k_D^{eff}\dfrac{\partial^2\ln c_e}{\partial x^2} + a_s F j_{li} = 0$	$\left.\dfrac{\partial\phi_e}{\partial x}\right\|_{x=0} = \left.\dfrac{\partial\phi_e}{\partial x}\right\|_{x=L} = 0$	(3.4)
Butler–Volmer equation	$j_{li} = i_0\left(\exp\left(\dfrac{\alpha_n F}{RT}\eta\right) - \exp\left(-\dfrac{\alpha_p F}{RT}\eta\right)\right)$		(3.5)
Exchange current density	$i_0 = Fk\left(c_e\right)^{\alpha_n}\left(c_{s,max}-c_{s,e}\right)^{\alpha_n}\left(c_{s,e}\right)^{\alpha_p}$		(3.6)
Overpotential	$\eta_i = \phi_{s,i} - \phi_{e,i} - U_{ocp,i}, i\in\{n,p\}$		(3.7)
Battery terminal voltage	$U(t) = \phi_s(L,t) - \phi_s(0,t) - R_f I$		(3.8)

TABLE 3.2
Nomenclature of parameters in Table 3.1

Parameter	Meaning	Parameter	Meaning	Parameter	Meaning
c_s	Lithium concentration in electrode	k^{eff}	Effective ionic conductivity	R	gas constant
c_e	Electrolyte concentration	k	Kinetic rate constant	I	Current
$c_{s,e}$	Lithium concentration at particle surface	η	Overpotential	R_f	Film resistance
$c_{s,max}$	Lithium maximum concentration	U	Battery terminal voltage	F	Faraday constant
D_s	Solid diffusion coefficient	U_{ocp}	Electrode open-circuit potential	σ^{eff}	Effective electrode conductivity
D_e^{eff}	Effective electrolyte diffusion coefficient	ϕ_s	Solid electrode potential	j_{li}	Lithium flux
ε_e	Volume fraction of electrolyte	ϕ_e	Electrolyte phase potential	n	Negative electrode
a_s	Specific interfacial area	i_0	Exchange current density	p	Positive electrode
α	Charge transfer coefficient	L_i	Length of positive or negative electrodes	A	Electrode area
r_s	Radius of active material particle	L_m	Length of the separator	T	Battery temperature
r	Radial coordinates in particles	L	Length of the battery	x	Electrode thickness

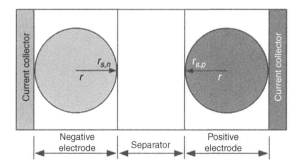

FIGURE 3.3   Schematic diagram of SP model.

$$\begin{cases} J_p = -\dfrac{I}{a_p FAL_p} \\[3mm] J_n = +\dfrac{I}{a_n FAL_n} \end{cases} \tag{3.1}$$

where J_i represents the average pore wall flux of the electrode, and a_i represents the specific surface area of the electrode.

By combining relevant equations from the P2D model, the expression for the output voltage of the SP model is given by Equation (3.2) [22].

$$U(t) = U_p\left(\frac{c_{s,p}}{c_{s,p,\max}}\right) - U_n\left(\frac{c_{s,n}}{c_{s,n,\max}}\right) + \eta_p - \eta_n - \frac{R_f}{A}I \tag{3.2}$$

Compared to the P2D model, the SP model has fewer parameters, which reduces the difficulty of parameter identification to some extent. However, under high current rate conditions, the simulation results of the SP model deviate significantly from the experimental measurements more than the P2D model [23].

3.2.3 ENHANCED SINGLE PARTICLE MODEL

To comprehensively consider the impact of electrode phase concentration and electrode phase potential on the battery output voltage, researchers have combined the electrolyte dynamics equations with the SP model to further develop the ESP model [24, 25]. This model addresses the limitations of the traditional SP model in terms of accuracy under high current rate and dynamic conditions, as shown in Figure 3.4.

By incorporating the electrode phase potential ($\varphi_{e,i}$) into the SP model, the output voltage expression of the ESP model can be derived, as shown in Equation (3.3) [26, 27].

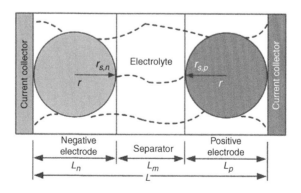

FIGURE 3.4   Schematic diagram of ESP model.

$$U(t) = U_p\left(\frac{c_{s,p}}{c_{s,p,\max}}\right) - U_n\left(\frac{c_{s,n}}{c_{s,n,\max}}\right) + \eta_p - \eta_n +$$

$$\phi_{e,p}(L) - \phi_{e,n}(0) - \frac{R_f}{A}I$$

(3.3)

The research focus of the ESP model lies in constructing the expression of the electrode phase potential and its corresponding solution form. In order to simplify the complexity of the equations and facilitate the solution process, it is necessary to reduce the order of the expression for the electrode phase potential. Such reduction methods can effectively reduce the model complexity and facilitate the implementation of the solution.

3.2.4 PHYSICS-BASED FRACTIONAL-ORDER MODEL

The impedance of LIBs manifests in various polarization characteristics, including ohmic polarization, activation polarization, charge transfer polarization, electrolyte diffusion polarization, and solid–phase diffusion polarization. Among these, for the study of battery dynamics and aging, the research and modeling of solid-phase diffusion polarization are particularly important [28]. The PBFOM is developed based on the SP model. It approximates the solid-phase diffusion mechanism using Padé approximation, obtaining a fractional-order model (FOM) for solid-phase diffusion. It combines this model with the medium-to-high frequency dynamic model to form a complete FOM of the battery, as shown in Figure 3.5 [29–31]. The equilibrium

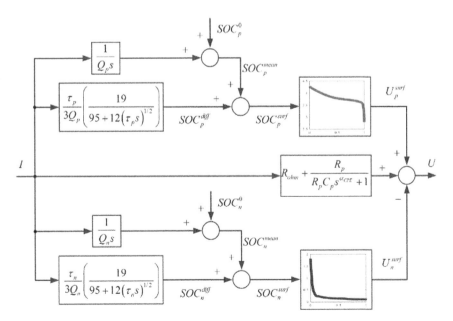

FIGURE 3.5 Schematic diagram of PBFOM.

potential curves of the positive and negative electrodes are obtained based on half-cell experiments and serve as benchmark data for this model.

The specific meanings of the parameters in PBFOM are presented in Table 3.3. The calculation process of SoC is represented by Equation (3.4).

$$
\begin{cases}
SOC_p^{surf} = SOC_p^0 + \dfrac{I}{Q_p s} + I \dfrac{\tau_p}{Q_p} \left(\dfrac{19}{95 + 12(\tau_n s)^{1/2}} \right) \\
SOC_n^{surf} = SOC_n^0 + \dfrac{I}{Q_n s} + I \dfrac{\tau_n}{Q_n} \left(\dfrac{19}{95 + 12(\tau_n s)^{1/2}} \right)
\end{cases}
\tag{3.4}
$$

The expressions for the capacity of the positive and negative electrodes are given by Equation (3.5). Here, δ_i represents the electrode thickness, and the meanings of other parameters are described in Table 3.2.

$$
\begin{cases}
Q_p = -\dfrac{A\delta_p \varepsilon_{s,p} F}{3600} c_{s,p,\max} \\
Q_n = -\dfrac{A\delta_n \varepsilon_{s,n} F}{3600} c_{s,n,\max}
\end{cases}
\tag{3.5}
$$

The identification of the medium-to-high frequency parameters in this model utilizes electrochemical impedance spectroscopy (EIS) of the LIB and considers factors such

---

TABLE 3.3
Nomenclature of parameters in Figure 3.5

Parameter	Meaning	Parameter	Meaning	Parameter	Meaning
i	Represents positive (p) or negative (n)	SOC_i^0	Initial SoC of the electrode	α_{CPE}	Order of the constant phase element
τ_i	Time constants for solid-phase diffusion	SOC_i^{mean}	Average SoC of the electrode	C_p	Internal polarization capacitance
U_i^{surf}	Electrode equilibrium potential curve	SOC_i^{diff}	Difference between the surface SOC and average SOC	R_p	Internal polarization resistance
Q_i	Electrode capacity	SOC_i^{surf}	Surface SOC	R_{ohm}	Ohmic resistance
s	Laplace variable ($s=j\omega$)				

as porous electrode dispersion effects. The constant phase element (CPE) is employed to replace the conventional capacitor for characterization purposes.

3.3   EQUIVALENT CIRCUIT MODELS

ECMs are empirical models used to describe the characteristics of LIBs. The structure of these models consists of components such as inductors, resistors, capacitors, voltage sources, and current sources. Model parameter values are obtained through specific parameter identification methods to describe the charge and discharge characteristics of the battery under specific conditions [32, 33]. ECMs utilize mathematical modeling techniques to numerically represent the model parameters and, in conjunction with appropriate state diagnosis algorithms, achieve the diagnosis and estimation of relevant battery states. Currently, research on ECMs is continuously evolving and includes models such as the Rint model [34–36], Thevenin model [37–40], second-order RC model [41–46], PNGV model [47, 48], and GNL model [49].

In the study of ECMs, three main aspects are emphasized:

- Firstly, researchers strive to improve the structure of ECMs to enhance its accuracy and applicability. This may involve adding more components or submodels to better describe the behavior of the battery. For example, introducing capacitor elements to consider the electrochemical polarization effect or voltage decay.
- Secondly, modeling methods for model parameters are also a key research area. Factors such as temperature, current, SOC, and health condition of the battery directly affect the parameters of ECMs. These parameters are considered non-linear functions in ECMs to reflect their relationship with the aforementioned factors. Therefore, researchers are devoted to developing precise modeling methods to obtain accurate numerical values for these parameters. This may involve techniques such as experimental testing, data fitting, and system identification.
- Thirdly, model-based battery state estimation is another important direction in the research of ECMs. By integrating real-time measurement data of the battery with ECMs, filtering, optimal estimation, and other algorithms can be used to infer the battery's state parameters such as SOC, internal resistance, and capacity decay. This contributes to monitoring and predicting the performance and lifespan of the battery.

3.3.1   RINT MODEL

The Rint model is a simple ECM, consisting of a voltage source and a DC internal resistance. Its structure is illustrated in Figure 3.6 [34–36]. In this model, U_{ocv} represents the open-circuit voltage, which is a function of SOC and temperature. R_{int} represents the DC internal resistance of the battery, which exhibits a certain degree of non-linear relationship with current, SOC, temperature, and aging level.

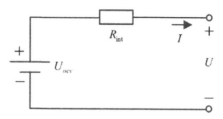

FIGURE 3.6   Rint model.

From Figure 3.6, it can be observed that the model parameters satisfy Equation (3.6).

$$U = U_{ocv} - IR_{int} \tag{3.6}$$

The Rint model has a simple structure and a small number of parameters. Its main focus of research lies in the mapping relationship between model parameters and factors such as battery temperature, current, SOC, and SOH. However, due to the lack of consideration for internal polarization effects within the battery, its applicability is relatively limited. It is typically used to describe battery characteristics under constant current conditions and is difficult to apply in dynamic operating scenarios.

3.3.2 THEVENIN MODEL

The Thevenin model, also known as the first-order RC model, is an extension of the Rint model that incorporates an RC network to describe the generation and elimination of battery polarization phenomena. This model is illustrated in Figure 3.7 [37–40]. In this model, R_o represents the ohmic resistance, R_p represents the polarization resistance, and C_p represents the polarization capacitance. By considering the effect of battery polarization, this model is capable of describing both the static and dynamic characteristics of the battery. Furthermore, the computation of this model is relatively simple, making it a widely applied ECM in engineering.

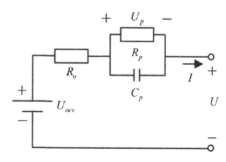

FIGURE 3.7   Thevenin model.

The output voltage of the Thevenin model can be expressed as in Equation (3.7):

$$\begin{cases} U = U_{ocv} - U_p - IR_o \\ U_p^k = IR_p\left[1 - \exp\left(-\dfrac{\Delta t}{R_p C_p}\right)\right] + U_p^{k-1}\exp\left(-\dfrac{\Delta t}{R_p C_p}\right) \end{cases} \tag{3.7}$$

To estimate the state of battery more accurately, researchers have made improvements to the Thevenin model to consider phenomena such as diffusion, hysteresis effects, and self-discharge processes. Wang et al. [50] proposed a modified Thevenin model with a diffusion resistor, as shown in Figure 3.8. This model incorporates modeling of the diffusion resistor (R_{diff}) based on extensive experimental data, to simulate the diffusion phenomenon in the low-frequency range, without affecting the calculation of peak discharge and charge currents.

Due to the hysteresis effects present in lithium iron phosphate batteries during charging and discharging processes, direct application of the Thevenin model for predicting battery state can lead to certain errors. Therefore, Wang et al. [51] proposed a Thevenin model that takes into account the hysteresis effects of the battery and combined it with the extended Kalman filter (EKF) method for battery power estimation, as shown in Figure 3.9. This improved model, by considering the hysteresis effects, can more accurately capture the dynamic behavior of the battery, thereby enhancing the precision of battery state estimation. The application of the EKF method further enhances the accuracy and robustness of battery power estimation, which is of significant importance for performance analysis and optimization control of lithium iron phosphate batteries.

As the number of battery cycles and calendar aging increase, the available capacity of the battery also undergoes changes. To address this issue, Dong et al. [52] proposed a Thevenin model that takes into account the self-discharge phenomenon. The model consists of two sub-models, namely the run time-based model and the voltage-current characteristics-based model. The run time-based model is composed of three components connected in parallel: the self-discharge resistor ($R_{self-discharge}$), the capacity (C_N), and the controlled current source (I_{bat}), as shown in Figure 3.10. This improved Thevenin model, by incorporating the consideration of self-discharge

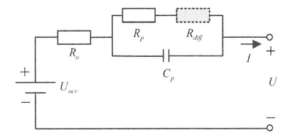

FIGURE 3.8 Thevenin model with diffusion resistance.

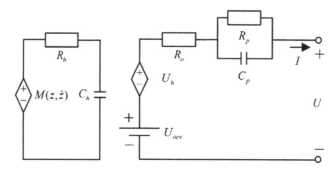

FIGURE 3.9 Thevenin model with hysteresis.

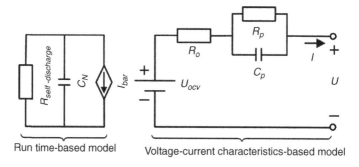

Run time-based model   Voltage-current characteristics-based model

FIGURE 3.10 Thevenin model considering self-discharge phenomenon.

phenomenon, can more accurately describe the capacity degradation behavior of the battery.

When estimating battery power and other states, there is often a trade-off between the complexity and accuracy of the battery model. To improve the accuracy of battery power estimation, Feng et al. [53] proposed a novel ECM, as shown in Figure 3.11. This model is based on the Thevenin model and introduces a moving average noise module to compensate for dynamic processes that the original model cannot describe, such as low-frequency diffusion phenomena.

3.3.3 SECOND-ORDER RC MODEL

The physical and chemical reactions in LIBs are highly complex and occur on different time scales. However, the Thevenin model only includes an RC network and cannot describe additional polarization phenomena that occur at different time scales. The second-order RC model [41–46] can be seen as an improved version of the Thevenin model, as it adds an additional RC network. These two RC networks describe the electrochemical polarization and concentration polarization phenomena during the charge and discharge processes of the battery, respectively, and hence the model is also referred to as the dual polarization (DP) model, as shown in Figure 3.12.

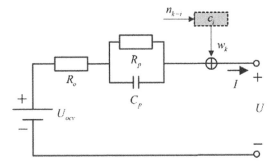

FIGURE 3.11   Thevenin model with a moving average noise.

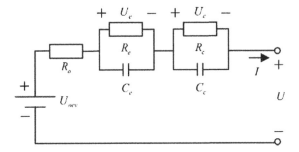

FIGURE 3.12   Second-order RC model.

In Figure 3.12, R_e represents the charge transfer impedance, C_e represents the charge transfer capacitance, R_c represents the concentration impedance, C_c represents the concentration capacitance, U_e represents the electrochemical polarization voltage, and U_c represents the concentration polarization voltage.

The discrete equation for battery voltage can be expressed as in Equation (3.8):

$$\begin{cases} U = U_{ocv} - U_e - U_c - IR_o \\ U_e^k = IR_e \left[1 - \exp\left(-\dfrac{\Delta t}{R_e C_e} \right) \right] + U_e^{k-1} \exp\left(-\dfrac{\Delta t}{R_e C_e} \right) \\ U_c^k = IR_c \left[1 - \exp\left(-\dfrac{\Delta t}{R_c C_c} \right) \right] + U_c^{k-1} \exp\left(-\dfrac{\Delta t}{R_c C_c} \right) \end{cases} \tag{3.8}$$

Compared to the Rint model and Thevenin model, the second-order RC model has a more complex structure. However, its advantage lies in its ability to simultaneously consider ohmic polarization, electrochemical polarization, and concentration polarization, thus providing a more accurate description of the dynamic polarization characteristics of the battery under high current rate conditions [54]. Therefore, the second-order RC model achieves higher accuracy in estimating the SOC, SOP, and SOE of the battery under dynamic operating conditions, making it widely adopted in practical applications.

However, as the number of RC networks in the ECMs increases, although it can improve the model accuracy to some extent, it also leads to excessive model complexity and significantly increases the computational time [55]. Therefore, in practical applications, it is not advisable to blindly increase the number of RC networks. Instead, the appropriate order of RC networks should be chosen based on the requirements to balance the accuracy and computational time.

3.3.4 PNGV MODEL

To comprehensively consider the polarization characteristics of batteries and improve the accuracy of the model, the partnership for a new generation of vehicles (PNGV) model was established by the United States Government and the United States Council for Automotive Research (USCAR) in 1993 [56]. The PNGV model is also an improvement upon the Thevenin model, incorporating a capacitor in series to describe the polarization phenomenon of the open-circuit voltage of the battery over time. The structure of this model is illustrated in Figure 3.13 [47, 48].

According to Figure 3.13, it can be observed that the voltage relationship of this model satisfies Equation (3.9):

$$
\begin{cases}
U = U_{ocv} - U_Q - U_p - IR_o \\
\dfrac{dU_Q}{dt} = \dfrac{I}{C_Q} \\
U_p^k = IR_p \left[1 - \exp\left(-\dfrac{\Delta t}{R_p C_p} \right) \right] + U_p^{k-1} \exp\left(-\dfrac{\Delta t}{R_p C_p} \right)
\end{cases}
\tag{3.9}
$$

Due to significant discrepancies between the impedance spectrum of the integer-order PNGV model and the EIS of actual batteries, especially in the low-frequency

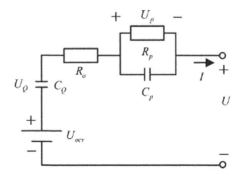

FIGURE 3.13 PNGV model.

range of impedance, Liu et al. [57] proposed a fractional-order PNGV model. This model replaces the integer-order capacitance with a fractional-order capacitance and combines it with the extended Kalman filtering method to accurately estimate the battery's SOC.

3.3.5 GNL MODEL

The GNL model can be considered as an extension of the second-order RC model by adding a capacitance in series or as an extension of the PNGV model by adding an RC network in series. The structure of this model is depicted in Figure 3.14 [49]. Consequently, the GNL model has a larger number of parameters, which makes model parameter identification and state estimation more complex. Additionally, the computation time required for the GNL model is also increased.

The expression satisfied by the GNL model is given by Equation (3.10):

$$
\begin{cases}
U = U_{ocv} - U_Q - U_e - U_c - IR_o \\
\dfrac{dU_Q}{dt} = \dfrac{I}{C_Q} \\
U_e^k = IR_e\left[1 - \exp\left(-\dfrac{\Delta t}{R_e C_e}\right)\right] + U_e^{k-1}\exp\left(-\dfrac{\Delta t}{R_e C_e}\right) \\
U_c^k = IR_c\left[1 - \exp\left(-\dfrac{\Delta t}{R_c C_c}\right)\right] + U_c^{k-1}\exp\left(-\dfrac{\Delta t}{R_c C_c}\right)
\end{cases}
\tag{3.10}
$$

3.3.6 FRACTIONAL-ORDER MODEL

The integer-order equivalent circuit models struggle to accurately describe the frequency-domain impedance of LIBs, especially in the low-frequency range [58, 59]. To address the discrepancy between the integer-order models and the EIS in the

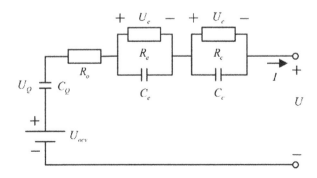

FIGURE 3.14 GNL model.

low-frequency range, researchers have proposed FOMs. The FOM is an improved model based on the EIS, capable of capturing the Warburg impedance element (Z_w) or CPE, and can be seen as an enhancement to the integer-order ECMs [60, 61]. The Warburg impedance element is used to characterize the low-frequency impedance behavior of the battery and its calculation is given by Equation (3.11) [59]. Here, R_d and τ represent the characteristic resistance and time constant, respectively. Vehicular Technology Society [62] extended the Rint model by introducing a Warburg impedance element, proposing a battery terminal voltage estimation method based on a FOM for SOP prediction, as shown in Figure 3.15.

$$Z_W(s) = R_d \frac{\tanh\sqrt{s\tau}}{\sqrt{s\tau}} \tag{3.11}$$

The voltage response of the Warburg impedance element can be decomposed into two parts: zero-input response and zero-state response. This decomposition helps simplify the non-linear modeling process of the Warburg impedance element. While this model can provide concise expressions, it sacrifices the non-linear characteristics of the model.

To better describe the polarization dynamics in batteries, Ye et al. [63] combined fractional-order calculus with the second-order RC model, replacing the capacitor element in the integer-order model with a CPE, as shown in Figure 3.16. This model can provide a more accurate description over a wide frequency range. The CPE impedance expression can be calculated using Equation (3.12) [64]. Here, Z_{CPE} represents the CPE impedance expression, C_{CPE} is the capacitance coefficient, and α_{CPE} represents the fractional order of the CPE ($0 \le \alpha_{CPE} \le 1$).

$$Z_{CPE}(s) = \frac{1}{C_{CPE}s^{\alpha_{CPE}}} \tag{3.12}$$

To further enhance the accuracy of the model, the combination of the Warburg impedance element and CPE can be applied to the ECM, as shown in Figure 3.17 [65]. Although this improvement increases the accuracy of the model, it also makes the model structure more complex and increases the computational burden. However, when the model order reaches the third order or higher, the increase in computational

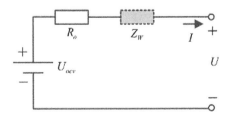

FIGURE 3.15   Simplified FOM.

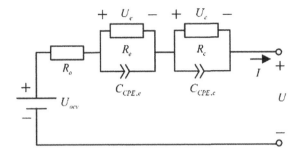

FIGURE 3.16 Second-order RC FOM.

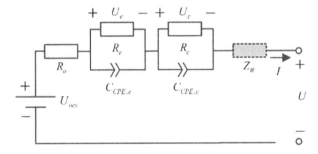

FIGURE 3.17 Improved second-order RC FOM.

burden far exceeds the rate of improvement in accuracy. Therefore, a higher model order does not necessarily lead to better performance [64, 66].

3.4 THERMAL MODELS

During the operation of LIBs, complex physical and chemical reactions occur internally, resulting in battery heating. These major heat-generating reactions in batteries involve the solid electrolyte interphase (SEI) film decomposition, electrolyte decomposition, positive electrode decomposition, reactions of deposited lithium metal with electrolyte, reactions of lithium-inserted carbon with solvent/binder, and heat generation from battery resistance [67]. When the battery undergoes multiple charge-discharge cycles, heat is generated internally. If the battery temperature becomes too high, it can lead to thermal runaway of the battery and thermal propagation within the battery pack. Therefore, it is necessary to study the thermal model of the battery and manage its operating conditions [68]. In this section, all the parameters used in the thermal models and their meanings are presented in Tables 3.1 and 3.4.

3.4.1 THERMAL EQUIVALENT CIRCUIT MODELS

In order to accurately predict the dynamic thermal characteristics of batteries and battery packs, researchers have proposed thermal equivalent circuit models [69–72].

TABLE 3.4
Nomenclature of parameters in TMs

Parameter	Meaning	Parameter	Meaning	Parameter	Meaning
Q	Heat generation	m_i	Mass of ith node	K_{ij}	Conductance
Q_{rea}	Reaction heat generation	M	Mess of battery	r_i	Reaction rate of the ith reaction
Q_{act}	Active heat generation	i_s	Solid-phase current	A^h	Heat exchange area
Q_{ohm}	Ohmic heat generation	i_e	Liquid phase current	ρ	Density
$Q_{k \to i}$	Heat generation source inside control volume of the node	T_t^i	Temperature of ith node at t instant	c_i^h	Specific heat of ith node
i	ith node of the battery	$T_{t+\Delta t}^i$	Temperature of ith node at $t+$ Δt instant	$k_{0,i}$	Reaction temperature constant
j	jth node of the battery	T_{ref}	Reference temperature	$k_{0,i}^{ref}$	Reaction temperature constant at reference temperature
ΔH_i^{avg}	Average entropy variation of ith node	T^l	Fluid node temperature	k^h	Thermal conductivity
$\bar{H}_j$	Molar entropy of jth node	T_{amb}	Ambient temperature	h_{ij}	Heat transfer coefficient with the fluid
$\bar{H}_j^{avg}$	Average molar entropy of jth node	R_j	Polarization resistance of jth RC	h	Heat transfer coefficient of the cooling medium
C_j	Ion concentration of jth node	R_j^0	Pre-exponential coefficient of R_j	$E_{a,i}^R$	Reaction activation energy
C_p^h	Heat capacity	R_{cj}	Equivalent thermal resistance of heat exchange	v	Volume
ε_s	Volume fraction for solid phase	R_{kj}	Equivalent thermal resistance within the cell	U_j	Polarization voltage of jth RC

The key to these models lies in using equivalent circuit models with thermal resistances to represent processes such as heat generation and transfer within the battery. Due to the structural differences among different battery types and battery systems, the structure of the thermal equivalent circuit models can also vary [73, 74]. Therefore, in this section, we take the thermal equivalent circuit model of a cylindrical cell as an example to introduce the design concept of the model, as shown in Figure 3.18 [75]. Number 1 is the internal temperature of the battery, numbers 2 to 4 represent the temperatures at different surface locations of the battery, and number 5 denotes the ambient temperature. The heat transfer processes between the internal part of the cell and different surface locations can be represented by equivalent thermal resistances, denoted as R_{kj}, while the heat exchange between the battery surface and the environment can be represented by equivalent thermal resistances, denoted as R_{cj}.

The heat generation equation of the battery can be represented by Equation (3.13) [76].

$$Q = I\left(U_{ocv} - U\right) - IT\frac{\partial U_{ocv}}{\partial T} - \sum_i \Delta H_i^{avg} r_i - \int \sum_j \left(\bar{H}_j - \bar{H}_j^{avg}\right)\frac{\partial C_j}{\partial t}\,dv \quad (3.13)$$

In this equation, the first term on the right-hand side represents the heat dissipation due to internal resistance, which is an irreversible heat generation. The second term indicates reversible entropy, and the third term denotes the heat generated or consumed by other chemical reactions in the battery. The last term represents the mixing heat, which is generated due to concentration gradients within the battery. The latter two terms in Equation (3.13) contribute only a small amount of heat and can be neglected in engineering applications without introducing significant errors [77]. The energy balance equation can be represented by Equation (3.14) [75].

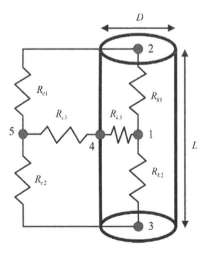

FIGURE 3.18 Thermal equivalent circuit model of cylindrical cell.

$$m_i c_i^{ch} \frac{T_{t+\Delta t}^i - T_t^i}{\Delta t} = \sum_j K_{ij} \left(T_{t+\Delta t}^j - T_{t+\Delta t}^i \right) + \sum_k Q_{k \to i} + \sum_j h_{ij} A_{ij}^h \left(T^l - T_{t+\Delta t}^i \right) \quad (3.14)$$

3.4.2 THERMAL-ELECTROCHEMICAL COUPLED MODELS

The P2D model describes the electrochemical processes of the battery and elaborates on the influence of temperature on model parameters. Therefore, the heat generation and distribution in the battery have also become one of the research focuses [17]. In the process of establishing a thermal model based on the P2D model, the heat sources consist of three parts: reversible entropy heat (Q_{rea}), irreversible electrochemical reaction heat (Q_{act}), and ohmic heat (Q_{ohm}) [78–80]. The energy balance equation and heat generation equation can be represented by Equations (3.15) and (3.16), respectively [80].

$$\rho C_p^h \frac{\partial T}{\partial t} = k^h \nabla^2 T + Q_{rea} + Q_{act} + Q_{ohm} \quad (3.15)$$

$$\begin{cases} Q_{rea} = jT \dfrac{3\varepsilon_s}{r_s} \dfrac{\partial U_{ocv}}{\partial T} \\[2mm] Q_{act} = j \dfrac{3\varepsilon_s}{r_s} \eta \\[2mm] Q_{ohm} = -i_s \nabla \phi_s - i_e \nabla \phi_e \end{cases} \quad (3.16)$$

The temperature of the battery has a significant impact on the rate constants of electrochemical reactions, and this influence can be expressed using the Arrhenius equation [80], as shown below:

$$k_{0,i} = k_{0,i}^{ref} \exp \left[\frac{E_{a,i}^R}{R} \left(\frac{1}{T_{ref}} - \frac{1}{T} \right) \right] \quad (3.17)$$

Due to the coupling between the parameters of the electrochemical model and the thermal model, they together form a thermo-electrochemical coupling model, which is used to describe the electrochemical dynamics and thermal characteristics of batteries.

3.4.3 THERMAL-ELECTRICAL COUPLED MODELS

The TMs can be coupled not only with EMs but also with ECMs [81]. When describing the heat generation in batteries, the Bernardi model [82, 83] calculates based on the electrical properties of the battery, and it has been widely applied in engineering practice. The Bernardi model simplifies the energy balance equation and the heat generation equation as follows:

$$MC_p^h \frac{dT}{dt} + hA^h(T - T_{amb}) = Q \tag{3.18}$$

$$Q = IT \frac{dU_{ocv}}{dT} + I\left(U_{ocv} - U\right) \tag{3.19}$$

To integrate TM with ECMs, the second term in Equation (3.19) can be expressed as the sum of the power dissipated by all internal resistances in the n-order RC ECM. Thus, Equation (3.19) can be rewritten as Equation (3.20):

$$Q = IT \frac{dU_{ocv}}{dT} + I^2 R_o + \sum_{j=1}^{n} \frac{U_j^2}{R_j} \tag{3.20}$$

Additionally, the temperature dependence of model parameters is commonly described using the Arrhenius equation, as shown in Equation (3.21) [82]:

$$R_j(T) = R_j^0 \exp\left(\frac{E_a^R}{RT}\right) \tag{3.21}$$

3.5 CONCLUSIONS

This chapter comprehensively discusses the applications of electrochemical models, equivalent circuit models, and thermal models in lithium-ion battery modeling and state estimation. Electrochemical models provide detailed descriptions of internal chemical reactions and characteristics of batteries, which are essential for a deep understanding of battery behavior but come with higher computational complexity. Equivalent circuit models simplify the model structure and computational complexity, making them suitable for battery state estimation and system design, but the model parameters lack explanations in terms of battery mechanisms. Thermal models consider the impact of internal temperature variations on battery performance and are generally coupled with electrochemical models or equivalent circuit models, providing important guidance for thermal management and life assessment. By integrating these models, comprehensive battery modeling, state estimation, and performance optimization can be achieved.

Future research directions include improving the accuracy and computational efficiency of the models, combining multiple model fusion methods to enhance the accuracy of battery state estimation, and exploring the application of emerging technologies such as deep learning in battery modeling and state estimation. These advancements will provide strong support for the development and application of battery technologies and drive sustainable development in areas such as electric transportation and renewable energy. Comprehensive studies indicate that electrochemical models, equivalent circuit models, and thermal models each have their advantages in battery modeling and state estimation. Selecting the appropriate model or combining multiple models will contribute to improving the accuracy and reliability of battery state estimation.

REFERENCES

[1] Teh, J.; Lai, C. M. Reliability Impacts of the Dynamic Thermal Rating and Battery Energy Storage Systems on Wind-Integrated Power Networks. *Sustain Energy Grids Networks*, **2019**, *20*, 100268. https://doi.org/10.1016/j.segan.2019.100268

[2] Chen, B.-H.; Chen, P.-T.; Yeh, Y. L.; Liao, H.-S. Establishment of Second-Order Equivalent Circuit Model for Bidirectional Voltage Regulator Converter: 48 V-Aluminum-Ion Battery Pack. *Energy Reports*, **2023**, *9*, 2629–2637. https://doi.org/10.1016/j.egyr.2023.01.086

[3] Metwaly, M. K.; Teh, J. Optimum Network Ageing and Battery Sizing for Improved Wind Penetration and Reliability. *IEEE Access*, **2020**, *8*, 118603–118611. https://doi.org/10.1109/ACCESS.2020.3005676

[4] Mohamad, F.; Teh, J.; Lai, C. M. Optimum Allocation of Battery Energy Storage Systems for Power Grid Enhanced with Solar Energy. *Energy*, **2021**, *223*, 12015. https://doi.org/10.1016/j.energy.2021.120105

[5] Metwaly, M. K.; Teh, J. Probabilistic Peak Demand Matching by Battery Energy Storage alongside Dynamic Thermal Ratings and Demand Response for Enhanced Network Reliability. *IEEE Access*, **2020**, *8*, 181547–181559. https://doi.org/10.1109/ACCESS.2020.3024846

[6] Lai, C. M.; Teh, J. Network Topology Optimisation Based on Dynamic Thermal Rating and Battery Storage Systems for Improved Wind Penetration and Reliability. *Appl Energy*, **2022**, *305*. https://doi.org/10.1016/j.apenergy.2021.117837

[7] Pai, H. Y.; Liu, Y. H.; Ye, S. P. Online Estimation of Lithium-Ion Battery Equivalent Circuit Model Parameters and State of Charge Using Time-Domain Assisted Decoupled Recursive Least Squares Technique. *J Energy Storage*, **2023**, *62*. https://doi.org/10.1016/j.est.2023.106901

[8] Lu, L.; Han, X.; Li, J.; Hua, J.; Ouyang, M. A Review on the Key Issues for Lithium-Ion Battery Management in Electric Vehicles. *J Power Sources*, **2013**, *226*, 272–288. https://doi.org/10.1016/j.jpowsour.2012.10.060

[9] Liu, F.; Shao, C.; Su, W.; Liu, Y. Online Joint Estimator of Key States for Battery Based on a New Equivalent Circuit Model. *J Energy Storage*, **2022**, *226*, 104780. https://doi.org/10.1016/j.est.2022.104780

[10] Xiao, Y.; Wen, J.; Yao, L.; Zheng, J.; Fang, Z.; Shen, Y. A Comprehensive Review of the Lithium-Ion Battery State of Health Prognosis Methods Combining Aging Mechanism Analysis. *J Energy Storage*, **2023**, *65*, 107347. https://doi.org/10.1016/j.est.2023.107347

[11] He, Z.; Yang, G.; Lu, L. A Parameter Identification Method for Dynamics of Lithium Iron Phosphate Batteries Based on Step-Change Current Curves and Constant Current Curves. *Energies (Basel)*, **2016**, *9* (6), 9060444. https://doi.org/10.3390/en9060444

[12] Han, S.; Tang, Y.; Khaleghi Rahimian, S. A Numerically Efficient Method of Solving the Full-Order Pseudo-2-Dimensional (P2D) Li-Ion Cell Model. *J Power Sources*, **2021**, *490*, 229571. https://doi.org/10.1016/j.jpowsour.2021.229571

[13] Li, W.; Fan, Y.; Ringbeck, F.; Jöst, D.; Han, X.; Ouyang, M.; Sauer, D. U. Electrochemical Model-Based State Estimation for Lithium-Ion Batteries with

Adaptive Unscented Kalman Filter. *J Power Sources*, **2020**, *476*, 228534. https://doi. org/10.1016/j.jpowsour.2020.228534

[14]   Liu, K.; Gao, Y.; Zhu, C.; Li, K.; Fei, M.; Peng, C.; Zhang, X.; Han, Q. L. Electrochemical Modeling and Parameterization towards Control-Oriented Management of Lithium-Ion Batteries. *Control Eng Pract*, **2022**, *124*. https://doi. org/10.1016/j.conengprac.2022.105176

[15]   Bizeray, A. M.; Zhao, S.; Duncan, S. R.; Howey, D. A. Lithium-Ion Battery Thermal-Electrochemical Model-Based State Estimation Using Orthogonal Collocation and a Modified Extended Kalman Filter. *J Power Sources*, **2015**, *296*, 400–412. https://doi. org/10.1016/j.jpowsour.2015.07.019

[16]   Forouzan, M. M.; Mazzeo, B. A.; Wheeler, D. R. Modeling the Effects of Electrode Microstructural Heterogeneities on Li-Ion Battery Performance and Lifetime. *J Electrochem Soc*, **2018**, *165* (10), A2127–A2144. https://doi.org/10.1149/2.128 1809jes

[17]   Xiao, Y.; Yang, F.; Gao, Z.; Liu, M.; Wang, J.; Kou, Z.; Lin, Y.; Li, Y.; Gao, L.; Chen, Y.; et al. Review of Mechanical Abuse Related Thermal Runaway Models of Lithium-Ion Batteries at Different Scales. *J Energy Storage*, **2023**, *64*, 107145 https://doi.org/ 10.1016/j.est.2023.107145

[18]   Li, C.; Cui, N.; Wang, C.; Zhang, C. Simplified Electrochemical Lithium-Ion Battery Model with Variable Solid-Phase Diffusion and Parameter Identification over Wide Temperature Range. *J Power Sources*, **2021**, *497*, 229900. https://doi.org/10.1016/ j.jpowsour.2021.229900

[19]   Andersson, M.; Streb, M.; Ko, J. Y.; Löfqvist Klass, V.; Klett, M.; Ekström, H.; Johansson, M.; Lindbergh, G. Parametrization of Physics-Based Battery Models from Input–Output Data: A Review of Methodology and Current Research. *J Power Sources*, **2022**, *521*, 230859. https://doi.org/10.1016/j.jpowsour.2021.230859

[20]   Guo, M.; Sikha, G.; White, R. E. Single-Particle Model for a Lithium-Ion Cell: Thermal Behavior. *J Electrochem Soc*, **2011**, *158* (2), A122. https://doi.org/ 10.1149/1.3521314

[21]   Khaleghi Rahimian, S.; Rayman, S.; White, R. E. Extension of Physics-Based Single Particle Model for Higher Charge-Discharge Rates. *J Power Sources*, **2013**, *224*, 180–194. https://doi.org/10.1016/j.jpowsour.2012.09.084

[22]   Cai, L.; White, R. E. Mathematical Modeling of a Lithium Ion Battery with Thermal Effects in COMSOL Inc. Multiphysics (MP) Software. *J Power Sources*, **2011**, *196* (14), 5985–5989. https://doi.org/10.1016/j.jpowsour.2011.03.017

[23]   Mehta, R.; Gupta, A. An Improved Single-Particle Model with Electrolyte Dynamics for High Current Applications of Lithium-Ion Cells. *Electrochim Acta*, **2021**, *389*, 138623. https://doi.org/10.1016/j.electacta.2021.138623

[24]   Schmidt, A. P.; Bitzer, M.; Imre, Á. W.; Guzzella, L. Experiment-Driven Electrochemical Modeling and Systematic Parameterization for a Lithium-Ion Battery Cell. *J Power Sources*, **2010**, *195* (15), 5071–5080. https://doi.org/10.1016/ j.jpowsour.2010.02.029

[25]   Tanim, T. R.; Rahn, C. D.; Wang, C. Y. State of Charge Estimation of a Lithium Ion Cell Based on a Temperature Dependent and Electrolyte Enhanced Single Particle Model. *Energy*, **2015**, *80*, 731–739. https://doi.org/10.1016/j.energy.2014.12.031

[26]   Prada, E.; Di Domenico, D.; Creff, Y.; Bernard, J.; Sauvant-Moynot, V.; Huet, F. Simplified Electrochemical and Thermal Model of LiFePO 4 -Graphite Li-Ion Batteries for Fast Charge Applications. *J Electrochem Soc*, **2012**, *159* (9), A1508–A1519. https://doi.org/10.1149/2.064209jes

[27] Luo, W.; Lyu, C.; Wang, L.; Zhang, L. A New Extension of Physics-Based Single
      Particle Model for Higher Charge-Discharge Rates. *J Power Sources*, **2013**, *241*,
      295–310. https://doi.org/10.1016/j.jpowsour.2013.04.129
[28] Ovejas, V. J.; Cuadras, A. State of Charge Dependency of the Overvoltage Generated
      in Commercial Li-Ion Cells. *J Power Sources*, **2019**, *418*, 176–185. https://doi.org/
      10.1016/j.jpowsour.2019.02.046
[29] Yang, X.; Guo, D.; Dong, L.; Yang, G. Estimation Method of Solid Phase Diffusion
      Time-Constant of Lithium-Ion Battery Based on Time-Domain Data of Two-
      Electrode Battery and Neural-Network. In *2020 IEEE 9th International Power
      Electronics and Motion Control Conference, IPEMC 2020 ECCE Asia*; Institute
      of Electrical and Electronics Engineers Inc., **2020**; pp 2500–2504. https://doi.org/
      10.1109/IPEMC-ECCEAsia48364.2020.9367667
[30] Guo, D.; Yang, G.; Feng, X.; Han, X.; Lu, L.; Ouyang, M. Physics-Based Fractional-
      Order Model with Simplified Solid Phase Diffusion of Lithium-Ion Battery. *J Energy
      Storage*, **2020**, *30*, 101404. https://doi.org/10.1016/j.est.2020.101404
[31] Guo, D.; Yang, G.; Han, X.; Feng, X.; Lu, L.; Ouyang, M. Parameter Identification
      of Fractional-Order Model with Transfer Learning for Aging Lithium-Ion Batteries.
      Int J Energy Res, **2021**, *45* (9), 12825–12837. https://doi.org/10.1002/er.6614
[32] Guo, R.; Shen, W. A Review of Equivalent Circuit Model Based Online State of
      Power Estimation for Lithium-Ion Batteries in Electric Vehicles. *Vehicles*, **2021**, *4*
      (1), 1–31. https://doi.org/10.3390/vehicles4010001
[33] Chen, Y.; Yang, G.; Liu, X.; He, Z. A Time-Efficient and Accurate Open Circuit
      Voltage Estimation Method for Lithium-Ion Batteries. *Energies (Basel)*, **2019**, *12*,
      12091803. https://doi.org/10.3390/en12091803
[34] Liu, W.; Yang, G.; Meng, D.; Wang, B.; Ma, L. *Lithium-Ion Battery Modeling
      Method Considering Temperature and Current*; **2022**; LNEE, Vol. 891, Springer
      Link. https://doi.org/10.1007/978-981-19-1532-1_124
[35] Liu, W.; Yang, G.; Meng, D.; Li, L.; Wang, B. Modeling Method of Lithium-Ion
      Battery Considering Commonly Used Constant Current Conditions. *Diangong Jishu
      Xuebao/Trans China Electrotech Soc*, **2021**, *36*, 210297. https://doi.org/10.19595/
      j.cnki.1000-6753.tces.210297
[36] Burgos-Mellado, C.; Orchard, M. E.; Kazerani, M.; Cárdenas, R.; Sáez, D. Particle-
      Filtering-Based Estimation of Maximum Available Power State in Lithium-Ion
      Batteries. *Appl Energy*, **2016**, *161*, 349–363. https://doi.org/10.1016/j.apene
      rgy.2015.09.092
[37] Waag, W.; Fleischer, C.; Sauer, D. U. Adaptive On-Line Prediction of the Available
      Power of Lithium-Ion Batteries. *J Power Sources*, **2013**, *242*, 548–559. https://doi.
      org/10.1016/j.jpowsour.2013.05.111
[38] Jiang, J.; Liu, S.; Ma, Z.; Wang, L. Y.; Wu, K. Butler-Volmer Equation-Based Model
      and Its Implementation on State of Power Prediction of High-Power Lithium Titanate
      Batteries Considering Temperature Effects. *Energy*, **2016**, *117*, 58–72. https://doi.
      org/10.1016/j.energy.2016.10.087
[39] Tang, X.; Wang, Y.; Yao, K.; He, Z.; Gao, F. Model Migration Based Battery Power
      Capability Evaluation Considering Uncertainties of Temperature and Aging. *J Power
      Sources*, **2019**, *440*, 227141. https://doi.org/10.1016/j.jpowsour.2019.227141
[40] Yang, L.; Cai, Y.; Yang, Y.; Deng, Z. Supervisory Long-Term Prediction of State of
      Available Power for Lithium-Ion Batteries in Electric Vehicles. *Appl Energy*, **2020**,
      257, 114006. https://doi.org/10.1016/j.apenergy.2019.114006
[41] Niri, M. F.; Dinh, T. Q.; Yu, T. F.; Marco, J.; Bui, T. M. N. State of Power Prediction
      for Lithium-Ion Batteries in Electric Vehicles via Wavelet-Markov Load Analysis.

IEEE Trans Intell Transport Syst, **2021**, *22* (9), 5833–5848. https://doi.org/10.1109/
TITS.2020.3028024

[42] Shen, P.; Ouyang, M.; Lu, L.; Li, J.; Feng, X. The Co-Estimation of State of
      Charge, State of Health, and State of Function for Lithium-Ion Batteries in Electric
      Vehicles. *IEEE Trans Veh Technol*, **2018**, *67* (1), 92–103. https://doi.org/10.1109/
      TVT.2017.2751613

[43] Shu, X.; Li, G.; Shen, J.; Lei, Z.; Chen, Z.; Liu, Y. An Adaptive Multi-State Estimation
      Algorithm for Lithium-Ion Batteries Incorporating Temperature Compensation.
      Energy, **2020**, *207*, 118262. https://doi.org/10.1016/j.energy.2020.118262

[44] Zou, C.; Klintberg, A.; Wei, Z.; Fridholm, B.; Wik, T.; Egardt, B. Power Capability
      Prediction for Lithium-Ion Batteries Using Economic Nonlinear Model Predictive
      Control. *J Power Sources*, **2018**, *396*, 580–589. https://doi.org/10.1016/j.jpows
      our.2018.06.034

[45] Nejad, S.; Gladwin, D. T. Online Battery State of Power Prediction Using PRBS and
      Extended Kalman Filter. *IEEE Trans Ind Electron*, **2020**, *67* (5), 3747–3755. https://
      doi.org/10.1109/TIE.2019.2921280

[46] Tan, Y.; Luo, M.; She, L.; Cui, X. Joint Estimation of Ternary Lithium-Ion Battery
      State of Charge and State of Power Based on Dual Polarization Model. *Int J
      Electrochem Sci*, **2020**, *15* (2), 1128–1147. https://doi.org/10.20964/2020.02.34

[47] Castano-Solis, S.; Serrano-Jimenez, D.; Fraile-Ardanuy, J.; Sanz-Feito, J. Hybrid
      Characterization Procedure of Li-Ion Battery Packs for Wide Frequency Range
      Dynamics Applications. *Elect Power Syst Res*, **2019**, *166*, 9–17. https://doi.org/
      10.1016/j.epsr.2018.09.017

[48] Wang, Q.; Wang, J.; Zhao, P.; Kang, J.; Yan, F.; Du, C. Correlation between the
      Model Accuracy and Model-Based SOC Estimation. *Electrochim Acta*, **2017**, *228*,
      146–159. https://doi.org/10.1016/j.electacta.2017.01.057

[49] Bašic, M.; Vukadinovic, D.; Višnjic, V.; Rakic, I. Dynamic Equivalent Circuit
      Models of Lead-Acid Batteries – A Performance Comparison. In *IFAC-
      PapersOnLine*; Elsevier B.V., **2022**; Vol. 55, pp 189–194. https://doi.org/10.1016/
      j.ifacol.2022.06.031

[50] Wang, S.; Verbrugge, M.; Wang, J. S.; Liu, P. Power Prediction from a Battery State
      Estimator That Incorporates Diffusion Resistance. *J Power Sources*, **2012**, *214*, 399–
      406. https://doi.org/10.1016/j.jpowsour.2012.04.070

[51] Wang, Y.; Pan, R.; Liu, C.; Chen, Z.; Ling, Q. Power Capability Evaluation for
      Lithium Iron Phosphate Batteries Based on Multi-Parameter Constraints Estimation.
      J Power Sources, **2018**, *374*, 12–23. https://doi.org/10.1016/j.jpowsour.2017.11.019

[52] Dong, G.; Wei, J.; Chen, Z. Kalman Filter for Onboard State of Charge Estimation
      and Peak Power Capability Analysis of Lithium-Ion Batteries. *J Power Sources*,
      2016, *328*, 615–626. https://doi.org/10.1016/j.jpowsour.2016.08.065

[53] Feng, T.; Yang, L.; Zhao, X.; Zhang, H.; Qiang, J. Online Identification of Lithium-
      Ion Battery Parameters Based on an Improved Equivalent-Circuit Model and Its
      Implementation on Battery State-of-Power Prediction. *J Power Sources*, **2015**, *281*,
      192–203. https://doi.org/10.1016/j.jpowsour.2015.01.154

[54] He, H.; Xiong, R.; Guo, H.; Li, S. Comparison Study on the Battery Models Used
      for the Energy Management of Batteries in Electric Vehicles. *Energy Conversion and
      Management*, **2012**, *64*, 113–121. https://doi.org/10.1016/j.enconman.2012.04.014

[55] Lai, X.; Gao, W.; Zheng, Y.; Ouyang, M.; Li, J.; Han, X.; Zhou, L. A Comparative
      Study of Global Optimization Methods for Parameter Identification of Different
      Equivalent Circuit Models for Li-Ion Batteries. *Electrochim Acta*, **2019**, *295*, 1057–
      1066. https://doi.org/10.1016/j.electacta.2018.11.134

[56]   Nelson, P.; Bloom, I.; Amine, K.; Henriksen, G. Design Modeling of Lithium-Ion Battery Performance. *Journal of Power Sources*, **2022**, *110*, 437–444.

[57]   Liu, C.; Liu, W.; Wang, L.; Hu, G.; Ma, L.; Ren, B. A New Method of Modeling and State of Charge Estimation of the Battery. *J Power Sources*, **2016**, *320*, 1–12. https://doi.org/10.1016/j.jpowsour.2016.03.112

[58]   Cruz-Manzo, S.; Greenwood, P. An Impedance Model Based on a Transmission Line Circuit and a Frequency Dispersion Warburg Component for the Study of EIS in Li-Ion Batteries. *J Electroanal Chem*, **2020**, *871*, 114305. https://doi.org/10.1016/j.jelechem.2020.114305

[59]   Moya, A. A. Low-Frequency Development Approximations to the Transmissive Warburg Diffusion Impedance. *J Energy Storage*, **2022**, *55*, 105632. https://doi.org/10.1016/j.est.2022.105632

[60]   Andre, D.; Meiler, M.; Steiner, K.; Walz, H.; Soczka-Guth, T.; Sauer, D. U. Characterization of High-Power Lithium-Ion Batteries by Electrochemical Impedance Spectroscopy. II: Modelling. *J Power Sources*, **2011**, *196* (12), 5349–5356. https://doi.org/10.1016/j.jpowsour.2010.07.071

[61]   Cruz-Manzo, S.; Greenwood, P. Analytical Transfer Function to Simulate the Dynamic Response of the Finite-Length Warburg Impedance in the Time-Domain. *J Energy Storage*, **2022**, *55*, 105529. https://doi.org/10.1016/j.est.2022.105529

[62]   Vehicular Technology Society; Institute of Electrical and Electronics Engineers. *2015 IEEE Vehicle Power and Propulsion Conference (VPPC): Proceedings*, 19–22 October 2015, IEEE, Montreal, Quebec. https://ieeexplore.ieee.org/xpl/conhome/7352212/proceeding

[63]   Ye, L.; Peng, D.; Xue, D.; Chen, S.; Shi, A. Co-Estimation of Lithium-Ion Battery State-of-Charge and State-of-Health Based on Fractional-Order Model. *J Energy Storage*, **2023**, *65*, 107225. https://doi.org/10.1016/j.est.2023.107225

[64]   Lai, X.; He, L.; Wang, S.; Zhou, L.; Zhang, Y.; Sun, T.; Zheng, Y. Co-Estimation of State of Charge and State of Power for Lithium-Ion Batteries Based on Fractional Variable-Order Model. *J Clean Prod*, **2020**, *255*, 120203. https://doi.org/10.1016/j.jclepro.2020.120203

[65]   Liu, C.; Hu, M.; Jin, G.; Xu, Y.; Zhai, J. State of Power Estimation of Lithium-Ion Battery Based on Fractional-Order Equivalent Circuit Model. *J Energy Storage*, **2021**, *41*, 102954. https://doi.org/10.1016/j.est.2021.102954

[66]   Farmann, A.; Sauer, D. U. Comparative Study of Reduced Order Equivalent Circuit Models for On-Board State-of-Available-Power Prediction of Lithium-Ion Batteries in Electric Vehicles. *Appl Energy*, **2018**, *225*, 1102–1122. https://doi.org/10.1016/j.apenergy.2018.05.066

[67]   Zhao, Y.; Zou, B.; Zhang, T.; Jiang, Z.; Ding, J.; Ding, Y. A Comprehensive Review of Composite Phase Change Material Based Thermal Management System for Lithium-Ion Batteries. *Renewable and Sustainable Energy Reviews*, **2022**, *2022*, 112667. https://doi.org/10.1016/j.rser.2022.112667

[68]   Mallick, S.; Gayen, D. Thermal Behaviour and Thermal Runaway Propagation in Lithium-Ion Battery Systems – A Critical Review. *J Energy Storage*, **2023**, *2023*, 106894. https://doi.org/10.1016/j.est.2023.106894

[69]   Xu, C.; Wang, H.; Jiang, F.; Feng, X.; Lu, L.; Jin, C.; Zhang, F.; Huang, W.; Zhang, M.; Ouyang, M. Modelling of Thermal Runaway Propagation in Lithium-Ion Battery Pack Using Reduced-Order Model. *Energy*, **2023**, *268*, 126646. https://doi.org/10.1016/j.energy.2023.126646

[70]   Wang, G.; Kong, D.; Ping, P.; He, X.; Lv, H.; Zhao, H.; Hong, W. Modeling Venting Behavior of Lithium-Ion Batteries during Thermal Runaway Propagation by

Coupling CFD and Thermal Resistance Network. *Appl Energy*, **2023**, *334*, 120660. https://doi.org/10.1016/j.apenergy.2023.120660

[71]   Ma, Y.; Cui, Y.; Mou, H.; Gao, J.; Chen, H. Core Temperature Estimation of Lithium-Ion Battery for EVs Using Kalman Filter. *Appl Therm Eng*, **2020**, *168*, 114816. https://doi.org/10.1016/j.applthermaleng.2019.114816

[72]   Dai, H.; Zhu, L.; Zhu, J.; Wei, X.; Sun, Z. Adaptive Kalman Filtering Based Internal Temperature Estimation with an Equivalent Electrical Network Thermal Model for Hard-Cased Batteries. *J Power Sources*, **2015**, *293*, 351–365. https://doi.org/10.1016/j.jpowsour.2015.05.087

[73]   Forgez, C.; Vinh Do, D.; Friedrich, G.; Morcrette, M.; Delacourt, C. Thermal Modeling of a Cylindrical LiFePO4/Graphite Lithium-Ion Battery. *J Power Sources*, **2010**, *195* (9), 2961–2968. https://doi.org/10.1016/j.jpowsour.2009.10.105

[74]   Lin, X.; Perez, H. E.; Mohan, S.; Siegel, J. B.; Stefanopoulou, A. G.; Ding, Y.; Castanier, M. P. A Lumped-Parameter Electro-Thermal Model for Cylindrical Batteries. *J Power Sources*, **2014**, *257*, 12–20. https://doi.org/10.1016/j.jpowsour.2014.01.097

[75]   Broatch, A.; Olmeda, P.; Margot, X.; Agizza, L. A Generalized Methodology for Lithium-Ion Cells Characterization and Lumped Electro-Thermal Modelling. *Appl Therm Eng*, **2022**, *217*, 119174. https://doi.org/10.1016/j.applthermaleng.2022.119174

[76]   Liu, G.; Ouyang, M.; Lu, L.; Li, J.; Han, X. Analysis of the Heat Generation of Lithium-Ion Battery during Charging and Discharging Considering Different Influencing Factors. *Journal of Thermal Analysis and Calorimetry*, **2014**, *116*, 1001–1010. https://doi.org/10.1007/s10973-013-3599-9

[77]   Nazari, A.; Farhad, S. Heat Generation in Lithium-Ion Batteries with Different Nominal Capacities and Chemistries. *Appl Therm Eng*, **2017**, *125*, 1501–1517. https://doi.org/10.1016/j.applthermaleng.2017.07.126

[78]   Wang, D.; Zhang, Q.; Huang, H.; Yang, B.; Dong, H.; Zhang, J. An Electrochemical–Thermal Model of Lithium-Ion Battery and State of Health Estimation. *J Energy Storage*, **2022**, *47*, 103528. https://doi.org/10.1016/j.est.2021.103528

[79]   Wang, D.; Huang, H.; Tang, Z.; Zhang, Q.; Yang, B.; Zhang, B. A Lithium-Ion Battery Electrochemical–Thermal Model for a Wide Temperature Range Applications. *Electrochim Acta*, **2020**, *362*, 137118. https://doi.org/10.1016/j.electacta.2020.137118

[80]   Ren, H.; Jia, L.; Dang, C.; Qi, Z. An Electrochemical-Thermal Coupling Model for Heat Generation Analysis of Prismatic Lithium Battery. *J Energy Storage*, **2022**, *50*, 104277. https://doi.org/10.1016/j.est.2022.104277

[81]   Mesbahi, T.; Sugrañes, R. B.; Bakri, R.; Bartholomeüs, P. Coupled Electro-Thermal Modeling of Lithium-Ion Batteries for Electric Vehicle Application. *J Energy Storage*, **2021**, *35*, 102260. https://doi.org/10.1016/j.est.2021.102260

[82]   Hariharan, K. S. A Coupled Nonlinear Equivalent Circuit-Thermal Model for Lithium Ion Cells. *J Power Sources*, **2013**, *227*, 171–176. https://doi.org/10.1016/j.jpowsour.2012.11.044

[83]   Hou, G.; Liu, X.; He, W.; Wang, C.; Zhang, J.; Zeng, X.; Li, Z.; Shao, D. An Equivalent Circuit Model for Battery Thermal Management System Using Phase Change Material and Liquid Cooling Coupling. *J Energy Storage*, **2022**, *55*, 105834. https://doi.org/10.1016/j.est.2022.105834

4 Economic Analysis and Optimal Sizing of Battery-Integrated Residential Systems

Anisa Emrani, Youssef Achour, Naoufel Ennemiri, and Asmae Berrada

4.1 INTRODUCTION

The deployment of renewable energy has recently experienced rapid growth due to the many benefits it provides. At the end of 2021, the installed capacity of renewable energy (RE) reached 4050 MW out of a total capacity of 10,743 MW, representing a 37.7 percent share of the energy sources [1]. This RE capacity is divided as follows: 827 MW in solar projects (20.41 percent of capacity in RE), 1423 MW in wind projects (35.13 percent), and 1800 MW in hydropower projects with more than 25 plants (44.44 percent) [2]. However, renewable energy sources face the challenge of unstable production due to their intermittent nature [2]. To solve this problem, one of the most effective solutions is the deployment of energy storage systems (ESS). Indeed, ESS play a key role in increasing the penetration of renewable energy sources and improving the reliability of energy systems by offsetting the imbalance between energy supply and demand [3]. There are several energy storage technologies used for stationary power applications, which can be classified as mechanical, electrochemical, electromagnetic, and thermal energy storage [3]. Given the wide variety of ESS, it is necessary to select the most appropriate type of energy storage for each application [4].

In residential applications, battery-integrated residential systems refer to energy systems that combine batteries with residential buildings to optimize energy usage, improve self-sufficiency, and enhance the integration of renewable energy sources. These systems typically consist of multiple components, including photovoltaic (PV) panels, wind turbines (WT), biodigesters (BD), and energy storage batteries. Battery energy storage systems (BESS) are increasingly being considered for various applications in modern power grids [5]. The declining cost of these systems has generated commercial and customer interest, particularly in the combination of residential PV systems, WT, and BD with battery storage, known as the PV-WT-BD-BSS system [6].

Numerous academic studies have examined the economic value of PV-WT-BD-BSS for different individual systems [5, 7]. The characteristics of electrochemical

60   DOI: 10.1201/9781003441236-4

or battery storage systems, namely, discharge length, depth of discharge, and cycle life, have a major impact on the economic performance indicators in a variety of applications [8, 9]. According to the application, authors in [10] evaluated the levelized cost of electricity (LCOE) of four types of battery storage and discovered a wide variety of LCOE. Time-shifting, T&D investments, and energy management are examples of energy applications that are less expensive than voltage and frequency regulation [10]. In energy applications, increased battery utilization and longer discharge times result in a larger electricity flow [10]. Cycle life, which determines investment and replacement costs, plays a major role in how well batteries perform in diverse applications [10]. The study demonstrates that Pb-A is economically competitive for a small-scale application, while Li-ion works effectively in applications requiring a lot of cycles and a high energy/power ratio [10]. The LCOE calculated in [11] varies between the different scenarios, mostly because many operating factors were taken into account depending on the application. In most applications, Li-ion batteries perform more efficiently than other batteries. Because of its high efficiency, Li-ion generates a greater profit, according to [12]. Li-ion batteries are anticipated to dominate the energy market by 2030 due to their long cycle life, high efficiency, and significant potential for capital cost reduction [8].

On the other hand, the main problems with Pb-A are its short lifetime and low efficiency. A VRFB performs similarly to a Li-ion. However, a VRFB is not appropriate for power applications due to the higher capital cost of big stacks and membrane areas. When the discharge time was increased from 1 to 8 hours, Zakeri and Syri noticed a decrease in the LCOE [13]. Because Na-S has a smaller capital cost, its rate of return (ROR) is higher. The LCOE was determined by [14] for several electrochemical ESSs. The authors claim that Pb-A's initial investment cost is the primary factor influencing its widespread application in the energy storage industry. Ni-Cd is superior to Pb-A in terms of lifetime and specific energy, but its greater investment cost is a drawback. With advances in materials and manufacturing, it is anticipated that the LCOE of batteries will drop. The development of mechanical ESSs may, therefore, be limited by this decline, as predicted by Schmidt [10]. By 2030, Li-ion would be competitive with a discharge duration of less than 4 hours and 300 cycles per year. According to Schmidt et al. [10], the VRFB would be competitive for a larger number of operating cycles and a longer discharge time. The round-trip efficiency and the life cycle are also variables that have the biggest effects on future LCOE estimations [10]. For instance, a VRFB could become more affordable than a Li-ion ESS by 2030 with a 16 percent increase in efficiency.

For the integration of battery storage in residential buildings, there are several online tools available free of charge that can analyze the benefits of specific BESS configurations considering variations in load and PV size [15]. Examining how to maximize the market benefit is another way to optimally size the hybrid RE system with the incorporation of energy storage systems. Microgrids are a notable example, where the overall benefits of operating in grid-connected mode are maximized and the total costs of operating in islanded mode are minimized [16]. The levelized operating costs of the energy storage system and the other operational components are included in the overall costs of the microgrid, while the total benefits are computed as the difference between the benefits from selling power and the total operating costs. More details on these formulations are provided in [16]. Other methods maximize the difference between electricity purchased from the grid and power sold to the grid for the hybrid system rather than looking at the entire costs and benefits [17], while other studies have focused only on the daily operational profitability using time-shifting of energy output to match energy spot prices successfully without taking lifetime running costs into account [18].

This chapter proposes the development of an optimization model using the Fmincon interior point method to determine the optimal size of a battery-integrated residential system composed of PV, wind turbine, biodigester, and battery storage. The objective is to minimize the overall system cost of energy while considering reliability constraints. Various techno-economic indicators such as loss of power supply probability (LPSP), life cycle cost (LCC), and cost of energy (COE) are considered to provide the optimal configuration of the hybrid battery-integrated residential system based on an economic analysis.

This chapter is organized as follows. In Section 4.2, the definition and mathematical modeling of the PV-WT-BD-BS components are presented, along with the economic optimal sizing methodology employed. Section 4.3 provides a description of the case study. The results and discussion are presented in Section 4.4. Finally, Section 4.5 concludes the chapter and outlines future perspectives.

4.2 MATERIALS AND METHODS

This section provides definitions of each component of the battery-integrated residential system and their modeling equations. It describes the approach and methodology used to evaluate the performance and interactions of the PV-WT-BD-BS system. The definitions of the components will establish a clear understanding of their roles and contributions to the overall system. Furthermore, the modeling equations will capture the dynamic behavior and characteristics of each system, allowing for an accurate representation of its functioning within the hybrid energy system. A schematic view of the considered battery-integrated residential system is presented in Figure 4.1.

4.2.1 Definition and Mathematical Modeling of the PV-WT-BD-BSS System

PV Modules
Photovoltaic solar energy is obtained from the energy of the sun's rays. This is why the photovoltaic panels that will harvest them are installed on the roofs with the best possible orientation. The goal is that they are exposed as much as possible to the radiation of the sun, collect the photons of the sun, and then produce electricity. The composition of solar panels is designed in such a way that the superposition of the layers, charged negatively or positively, produces an electrical voltage when a photon passes through them. A wire connected to a positive terminal and another to the negative terminal, similar to a battery, allows the use of the energy thus produced [19]. Solar panels differ in the quality of silicon used in the manufacture of photovoltaic cells.

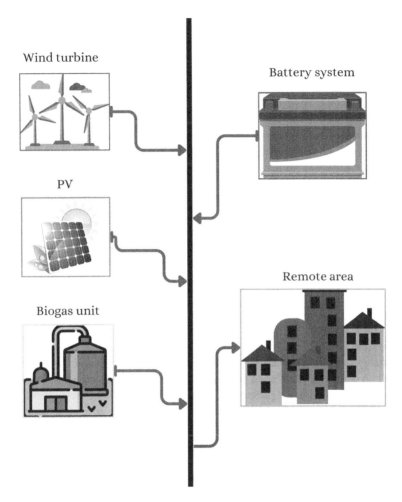

Wind turbine

Battery system

PV

Remote area

Biogas unit

FIGURE 4.1 Schematic view of the battery-integrated residential system.

Indeed, the monocrystalline panel's cells are composed of a single silicon block, which is the most expensive of photovoltaic panels and could be distinguished by its black color. It also produces the best efficiency, between 16 and 24 percent [19]. Polycrystalline panels are recognized by their dark blue color and are manufactured with several silicon crystals. Its efficiency ranges between 14 and 18 percent [19].

The amorphous panel can create electricity with artificial light. Its efficiency is between 5 and 7 percent. This low performance is due to the fact that its photovoltaic cells contain a smaller amount of silicon. But thanks to its flexibility, it is adaptable to supports with reduced surfaces [19]. The PV effect used in solar cells makes it possible to directly convert the light energy of solar rays into electricity through the production and transport of energy from photons to electrons in a semi-material conductor of positive and negative electrical charges under the effect of light. It is a semiconductor device that is able to conduct electricity and transform electrical energy from the light energy provided by an inexhaustible source of energy, the sun.

In this study, PV panels from Mitsubishi Electric (Model: PV-MLT260HC) were used with a length, width, and thickness of 1625 mm, 1019 mm, and 46 mm, respectively. The PV panels have a maximum power of 260 W. The power produced by the PV system depends on the overall irradiation and ambient temperature; it is obtained by Equations (4.1–4.2), where N_{pv} is the number of PV, G is the global solar irradiation of the site, η_{pv} is the efficiency of a photovoltaic module, T_{cref} is the reference temperature of the solar cell in standard operating conditions, G_{ref} is the solar radiation under standard test conditions, βt is the cell temperature coefficient, and NOCT is the nominal cell operating temperature.

$$P_{pv} = N_{pv} \times P_{pv\_rate} \times \eta_{pv} \times \frac{G}{G_{ref}} \times \left[1 - \beta_t \left(T_c - T_{cref}\right)\right]$$
(4.1)

$$T_c = T_a + \frac{(NOCT - 20)}{800} G$$
(4.2)

Wind System

Wind energy is the kinetic energy of the air masses moving around the globe. Wind energy is an indirect form of solar energy. Indeed, solar rays absorbed into the atmosphere cause temperature and pressure differences. This causes the air masses to move and accumulate kinetic energy. The wind turbine is coupled to an electric generator to create direct or alternating current. The generator is linked to a power grid or operates within a "stand-alone" system with a backup generator (for example, a generator set), battery, or other energy storage device [20].

There are two types of possible installations of wind turbines; industrial and domestic. In industrial installations, huge wind farms are connected to the power grid, while in domestic usage, small wind turbines are installed for residential applications. In reality, a wind turbine generates four times as much energy with a doubled blade area and eight times that amount with a doubled wind speed. Air density also affects power output: if the air is 10 °C cooler for a certain wind speed, a wind turbine will generate 3 percent more electricity. The key factors affecting wind power are the wind's speed and fluctuations. As a result, wind energy is unpredictable and intermittent [20]. It is important to note that the wind is more powerful and consistent at sea.

A rudder mechanism is used to direct the blade/rotor assembly into the wind. The majority of wind turbines begin operation when the wind speed is about 3 m/s and cease when the wind speed is 25 m/s. Typically, wind turbines are configured to maximize the potential of winds of moderate power [20].

The power produced by the wind system is calculated using Equation (4.3). P_w power generation by a wind turbine starts at a wind speed v higher than the start speed $v_{(cut-in)}$. The wind turbine produces a nominal power $P_{(wind\ rate)}$ after reaching the nominal speed v_{rate}. For safety reasons, the wind turbine stops at the cut-out speed.

$$P_w = \begin{cases} 0 & v < v_{cut-in} \\ \dfrac{N_{wt} \times P_{wind\_rate} \times \left(v^3 - v_{cut-in}^3\right)}{v_{rate}^3 - v_{cut-in}^3} & v_{cut\,in} < v < v_{rate} \\ N_{wt} \times P_{wind_{rate}} & v_{rate} \leq v \leq v_{cut-out} \\ 0 & v_{cut-out} \leq v \end{cases} \qquad (4.3)$$

Biodigester

Biodigesters are systems that use organic waste, such as food waste or agricultural residues, to produce biogas through a process called anaerobic digestion. Biogas can be used for heating, cooking, or generating electricity, providing an additional renewable energy source for residential systems.

The principle of biogas production is to place organic matter in a hermetically sealed, heated, and brewed tank known as a digester or bioreactor. Anaerobic bacteria feed on organic matter to multiply and thus produce biogas [11]. The latter allows the production of heat and/or electricity.

Equation (4.4) is used to calculate the electrical power produced by the biogas system, regardless of the type of waste. V_{biogas} is the volume of biogas, P_{cal} is the calorific value of the waste, and $\eta_{él}$ is the electrical efficiency of the unit.

$$P_{bio} = V_{biogaz} \times P_{cal} \times \eta_{a} \qquad (4.4)$$

Battery Storage System

Battery energy storage systems work under highly unique circumstances that vary from one technology to another. Typically, battery energy storage systems are identified by their state of charge (SOC). The starting SOC and the charge or discharge time of ESS are also critical parameters. Moreover, the majority of storage solutions are not 100 percent efficient. Indeed, losses happen throughout the charging and draining processes as well as at storage times. Therefore, the efficiency and self-discharge rate of the ESSs affect their SOC. The SOC depends on the preceding SOC state, as well as the energy used and produced between time t and time $t-1$. Depending on the load demand at a particular moment, global power generation might or might not be adequate to provide the necessary electricity. The time step is usually one hour. The expression of the SCO is given by Equations (4.5–4.6) [21].

$$Charge: SOC(t) = SOC(t-1) * (1-\sigma) + \left(P^g(t) - \frac{P^l(t)}{\mu_{inv}^*(t)} \right) * \mu_{inv} \qquad (4.5)$$

$$Discharge: SOC(t) = SOC(t-1) * (1-\sigma) + \left(\frac{P^l(t)}{\mu_{inv}^*(t)} - P^g(t) \right) / \mu_{bf} \qquad (4.6)$$

Lead-acid batteries are widely used as battery-integrated residential systems due to their favorable characteristics, such as a notable depth of discharge and high cycling stability [5]. In this study, the specific type of battery employed is the Sonnenschein A600 OPzV lead-acid battery, specifically designed for medium to large-scale applications. The selected model has a rated capacity of 20.192 kWh, a maximum depth of discharge (DOD) of 60 percent, a round-trip efficiency of 75 percent, and a service life of ten years. The maximum energy storage capacity of the battery bank, denoted as $E_{Batt-\ charg,max,}$ is determined by Equation (4.7). This equation considers the number of batteries (*NBAT*) and their individual capacity (*EBAT*), as well as the battery's capacity at a specific hour (*i*).

$$EBatt - charg, max = NBAT \times EBAT - Ebattery(i) \qquad (4.7)$$

Regarding the maximum discharged electricity, it can be mathematically represented as Equation (4.8) [5].

$$EBatt - disch, max = \eta bat \times (Ebattery(i) - (1 - DOD) \times EBAT) \qquad (4.8)$$

4.2.2 SYSTEM FINANCIAL INDICATORS

The financial viability of the battery residential hybrid system is a major factor in deciding the size of a battery energy storage system and the overall design of a hybrid RE system. Financial indicators have the advantage of providing a common decision-making unit, which makes it possible to compare various options. There are several distinct indicators that may be employed as optimizable design parameters. Numerous studies have examined the RE hybrid system's cost using LCC analysis. It entails calculating the net present value of all the costs anticipated to be incurred during the system's lifetime. It could be formulated as in Equation (4.9).

$$LCC = C_{Cap}^{npv} + C_{O\&M}{}^{npv} + C^{npv}{}_{R} + C^{npv}{}_{S} \qquad (4.9)$$

The total costs and revenues of the battery energy storage system throughout its operating lifetime are also evaluated through the calculation of other indicators such as the net present value (NPV) [20], which should be maximized, or the levelized cost of electricity (LCOE), and the levelized cost of storage (LCOS), which should be minimized [20].

4.2.3 ENERGY MANAGEMENT STRATEGY

The energy management strategy of the battery residential system, consisting of PV panels, wind turbines, a biodigester, and battery storage, effectively responds to the load demand in the residential building. This strategy involves optimizing the utilization of renewable energy sources and storage to ensure a reliable and sustainable power supply. The system operates by continuously monitoring the energy generation from the PV panels, wind turbines, and biodigester, as well as the energy consumption of the residential load. The energy management system intelligently manages the allocation and distribution of energy based on various factors, such as real-time energy production, battery storage capacity, and load demand.

When renewable energy sources produce an excess of electricity beyond the immediate demand, the surplus energy is stored in the battery storage system. The battery acts as a buffer, capturing and storing the excess energy for later use when the renewable energy generation is insufficient to meet the load demand. This helps in achieving a balanced energy supply and demand profile. During periods of high energy demand or when renewable energy generation is insufficient, the stored energy in the battery is discharged to supplement the load requirements. This ensures that the residential load is adequately, powered even in unfavorable conditions. This approach helps in achieving energy self-sufficiency, reducing reliance on the grid, and promoting sustainable energy consumption. Figure 4.2 is a flow chart of the energy management strategy used for the PV-WT-biogas-battery system to meet the residential energy demand.

FIGURE 4.2 Flow chart of the energy management strategy used for the PV-WT-BD-BSS to meet the residential energy demand.

4.2.4 OPTIMIZATION PROBLEM

Objective Function and Economic Analysis
The objective function considers minimizing the COE expressed in Equation (4.12). The COE calculation is based on the LCC analysis, which estimates the total cost of the project, as in Equations (4.10–4.11). To calculate the LCC, the capital costs of each component of the system (i.e., PV, wind turbines, biodigester, and battery storage) are taken into account. The LCC also takes into consideration operating and maintenance costs (calculated as a percentage of capital cost). Table 4.1 details the cost of each component. It should be noted that a discount rate (r) of 6 percent is used to calculate the net present value of the system over its life L_i.

$$C_{total-i} = CRF(r, L_i) \times (C_{inv-i} + C_{O\&M-i})$$

(4.10)

$$CRF(r, L_i) = \frac{r(1+r)^{L_i}}{(1+r)^{L_{i-1}}}$$

(4.11)

$$COE = \frac{\sum C_{total-i}}{E_{load}}$$

(4.12)

Design Variables of the Problem
The design variables that have been set for this study are the number of PV, the number of wind turbines, the volume of biogas to be produced, and the capacity of the battery bank. The developed model is able to determine the values of these variables for optimal and cost-effective operation.

Reliability Constraint
The proposed hybrid systems should meet the reliability constraint; therefore, the proposed objective function should be minimized while respecting a tolerable LPSP. The LPSP is an index that defines the probability of defective operation of the hybrid system when the load requirement is not fully met. The LPSP is calculated using Equation (4.13).

TABLE 4.1
System cost

Component	PV	WT	Biogas	Battery
Lifetime	25	25	25	25
C_{inv-i} (£)	2000	3200	1000	0.791 £/kWh
$C_{O\&M-i}$ (£) (% of C_{inv-i})	1%	3%	1%	1%

$$LPSP = \frac{\sum_{i=1}^{168}\left(E_{load}(t_i)-\left(E_{pv}(t_i)+E_{wind}(t_i)+E_{BioG}(t_i)+E_{ges\ decharge}(t_i)\right)\right)}{\sum_{i=1}^{168}E_{load}(t_i)} \quad (4.13)$$

4.2.5 Optimization Algorithm

The proposed problem formulation presents a non-linear constraint optimization problem that aims to minimize the system COE (objective function) and ensure the required percentage of LPSP (constraint). To solve the optimization problem, an Fmincon non-linear programming solver was implemented in the MATLAB optimization toolkit. Initially, the mathematical models of the plant components and the operational strategy were developed using MATLAB. The Fmincon solver is then called to find the minimum value of the multivariable constraint function defined using the interior point method. Optimization is completed when the maximum relative constraint violation is below the constraint tolerance and when the relative optimality measurement is below the optimal tolerance.

4.3 CASE STUDY-RELATED DATA

This case study consists of a residence located in Rabat (Figure 4.3). It is a "green" home producing electricity only through renewable and off-grid energy. Our residential system requires wind turbines, solar panels, and biogas as renewable energy systems. When electricity production exceeds demand, energy is stored in batteries.

Figure 4.4 shows the energy demand of the residential building considered. It reaches a maximum value of 7000 W and exhibits the same pattern each day. In our case, we have considered the simulation over a period of one week.

For the chosen location, the climatic data over the study period have been captured, which includes the ambient temperature, the solar radiation, and the wind speed temperature. According to Figure 4.5 (Ta=f[time]), the ambient temperature

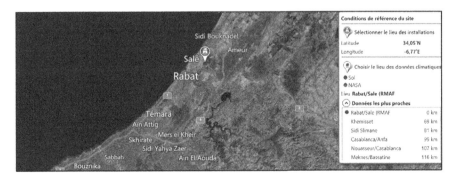

FIGURE 4.3 Location of residence.

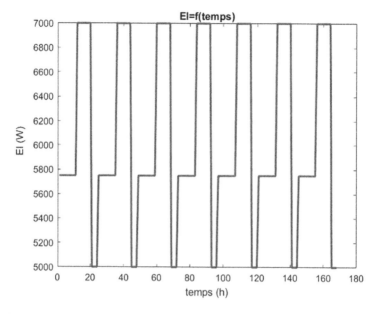

FIGURE 4.4   Energy demand over the study period (7 days).

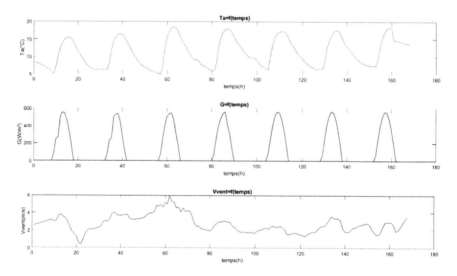

FIGURE 4.5   Climatic data.

drops at night and increases to its maximum during the day. G=f[time] shows that the irradiation is maximum during the day and absent during the night. The wind speed varies in a non-uniform way and reaches its maximum in the sixtieth hour of the week.

4.4 RESULTS AND DISCUSSION

In this section, the optimal hybrid system design results are first presented and discussed. The developed optimization model was simulated using MATLAB 2019a to evaluate the optimal configuration of the battery-integrated residential system. Tables 4.2–4.4 present the technical parameters used for the installation.

Figure 4.6 presents the progress of the optimization process to reach the optimal values of the design variables and thus the minimum cost while meeting the requirement to cover 100 percent of the demand without loss of power supply (LPSP = 0%).

The algorithm seeks the optimal sizing that will cover the energy demand, which explains the increase in COE due to the increase in the design values proposed by the algorithm during the first iterations. From the twenty-second iteration, the system

TABLE 4.2
Data used for PV

Parameter	Value
Efficiency	0.157
Power	260 W
PV module tilt angle	15°
Azimuth angle of PV modules	180°
Temperature coefficient	0.45%

TABLE 4.3
Data used for wind turbine

Parameter	Value
Height of wind turbine hub	10 m
Rated power	3000 W
Rotor diameter	12.8 m
Cut-in velocity	3 m/s
Rated velocity	9 m/s
Cut-out velocity	12 m/s

TABLE 4.4
Data used for biogas

Parameter	Value
Calorific value	21,524 kJ/m³
Efficiency	39%

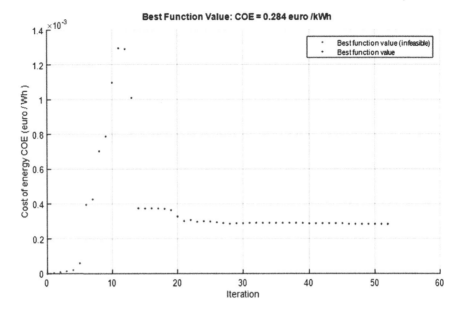

FIGURE 4.6   Progress of the optimization process to reach optimal values.

TABLE 4.5
Optimal values of design variables

Design variable	Optimal value
Number of PVs	12
Number of wind turbines	3
Volume of biodigester	2.55 m³
Battery capacity	4.34165 kWh

reaches the optimal values, which explains the stability of the graph. We also note that the first eight iterations give non-optimal values that do not satisfy the demand. From the eighth iteration to the twelfth, although the system satisfies the demand, it is oversized, which is not optimal since it increases the cost. The minimum cost obtained from this installation is 0.284 euro/kWh.

The values of the optimal design variables obtained are presented in Table 4.5. It is noted that the number of PVs is greater than the number of wind turbines, and that the volume of biogas required is very significant. This can be justified by the considered costs of each component. Indeed, the source that has the lowest cost was favored by the optimization algorithm in order to have the lowest value of the objective function (COE). The battery capacity is also logical, as we are on a domestic scale.

Figure 4.7 shows the hourly output of the PV modules during the study period. We note that the power produced by photovoltaic panels varies in a sinusoidal way. This

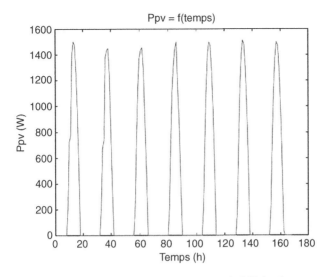

FIGURE 4.7 Production of PV modules over the study period (7 days).

can be explained by the fact that photovoltaic production depends on solar irradiation. The maximum output of PVs is 1.517 kW. The production of PVs is almost the same during each day (24 h), it reaches its maximum around noon, when the solar irradiation is maximum, and it is absent during the night.

Figure 4.8 presents the production of wind turbines during the studied period. The production of energy by wind turbines varies in a non-uniform way depending on the wind speed. The maximum value reached is 2.327 kW knowing that it becomes zero in cases where the wind speed is less than the speed of engagement, or greater than or equal to the speed of interruption.

Figure 4.9 shows that biogas electricity production is too high compared to other renewable energy sources. This is due to the large volume of biogas needed to cover enteric demand at the lowest cost in comparison with the number of PVs and wind turbines. Biogas production is the same every hour and equals 5.9225 kW.

In order to assess the capacity of the optimal design proposed to meet the energy demand of the system, Figure 4.10 depicts the demand curve and the curve that represents the sum of the renewable production in addition to the energy discharged from the battery bank when needed.

The curve representing production exceeds the curve representing demand. Therefore, the total energy supplied by renewable resources with the optimal size obtained is able to meet 100 percent of the energy demand and cover the needs of our residence without using the network.

To get a closer view of the energy flow, we have represented the energy charged and discharged from the battery system with renewable production and demand, as illustrated in Figure 4.11. The optimal battery for this application has a maximum capacity not exceeding 5 kWh. This is why their contribution is minimal compared to the production of renewable energies. However, the total satisfaction of the energy

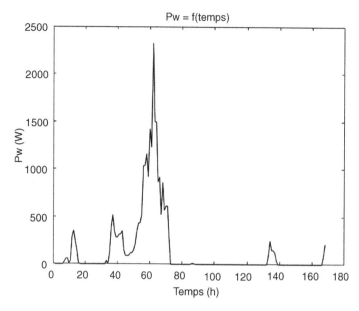

FIGURE 4.8   Production of WT over the study period (7 days).

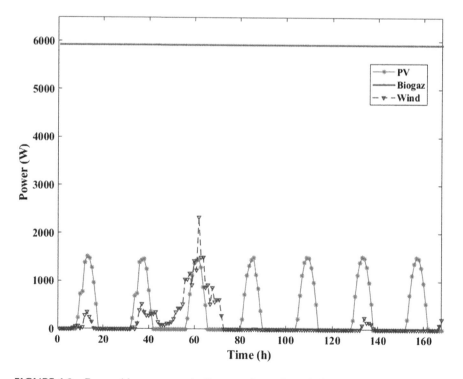

FIGURE 4.9   Renewable energy production over the study period.

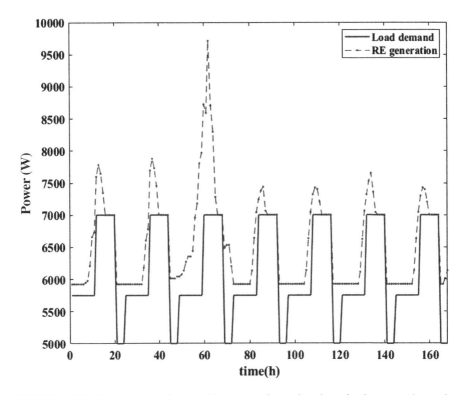

FIGURE 4.10 Comparison of renewable energy demand and production over the study period.

requirement can only be achieved by deploying it. As shown in Figure 4.11, the battery charges energy (green curve) when there is excess production (the red curve exceeds the blue curve) until saturation. Otherwise, the energy is discharged (purple curve).

Figure 4.12 shows the state of charge of the battery system over the study period. The simulation was started with 50 percent charged battery, as renewable energy production in the first few hours of the simulation was zero. When we have an excess of energy production, the battery stores the remaining energy. Otherwise, when we have an energy deficit, the battery is discharged to cover the needs of the residence. Note that the battery discharge will always take place towards the end of the day, when the energy production is very minimal, while it reaches 100 percent of its load around noon, when the production is at its maximum.

The comparison between the two energy curves provided without and with the battery system shows the positive contribution of the battery energy storage system to the installation. Indeed, the elimination of battery makes it difficult to meet 100 percent of energy demand, especially during the last hours of the day (at night), as shown in Figure 4.13. The battery is thus able to reduce energy losses by storing a maximum capacity, which in our case reaches 4.34 kWh, and reproducing it when needed. This also has a positive impact on the system's cost. Table 4.6 presents the costs required for the overall installation.

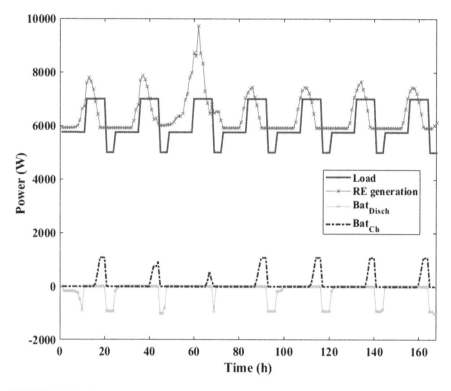

FIGURE 4.11   Optimal hybrid system energy exchange.

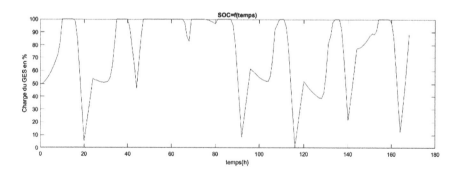

FIGURE 4.12   SOC of battery system over the study period.

4.5   CONCLUSION

In order to achieve sustainable and effective energy solutions, it is crucial to combine battery-integrated residential systems made up of a variety of energetic systems, including photovoltaic (PV) panels, wind turbines, biodigesters, and batteries. This integrated strategy enables the use of renewable energy sources, the

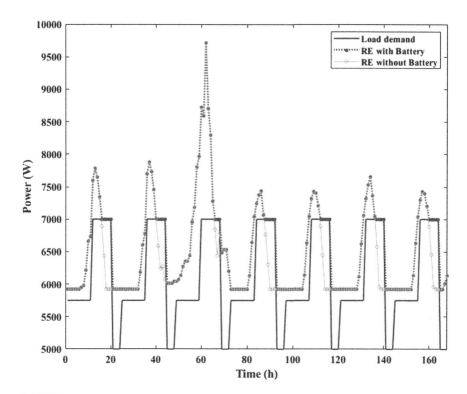

FIGURE 4.13 Battery energy storage system contribution to hybrid system operation.

TABLE 4.6
Different residential installation costs

Component	PV	WT	BD	Battery
$Cout_{Capital}(£)$	24,000	9600	5,922,521	4169
$C_{total}(£)$	466,253			
$C_{O\&M}(£)$	59,795			
$Cout_{installation.tot}(£)$	526,050			
$COE(£/kWh)$	0.3			

storage of extra energy, and the best possible resource usage, eventually leading to a more sustainable and resilient future. The optimal sizing of these systems, which refers to figuring out the proper capacity of each component to provide optimal economic performance metrics, is a vital issue. As part of the sizing process, variables including energy consumption, resource availability, and cost concerns are taken into account to strike the correct balance between efficacy, cost, and environmental impact.

This study explores the optimal design of hybrid renewable energy systems coupled with energy storage. The type of energy storage technology considered is battery storage. The main objective of this study is to provide an overview of the economic profitability and operation of the battery-integrated residential hybrid system, taking into account multiple renewable energy systems, including PV, wind turbines and biodigester with battery storage. A model with reliability constraints was used to identify the best design of the hybrid system for optimal operation while minimizing the cost of energy. Important results have been obtained which emphasize the benefits of this approach. Outputs from the design model include the number of PV modules and wind turbines, the volume of the biodigester, and the battery capacity. The optimal sizing of the battery-integrated residential system resulted in several notable advantages. Firstly, it maximized the utilization of renewable energy sources by efficiently capturing and storing energy from PV panels and wind turbines, reducing dependence on traditional energy sources, and lowering carbon emissions. Moreover, the integration of biodigesters into the system provided an additional renewable energy source through the conversion of organic waste into biogas. This not only reduced waste management challenges but also enhanced the overall sustainability of the system. Furthermore, the inclusion of batteries played a pivotal role in the system's economic performance metrics. They enabled the storage of surplus energy during periods of high production and released it during peak demand periods, thereby minimizing reliance on the grid and optimizing energy usage. This led to significant cost savings on electricity bills and improved the overall economic viability of the system.

In future work, ecological indicators such as CO_2 mitigation could be evaluated to provide a comprehensive analysis. In addition, the revenues generated by the energy systems studied could be included in the optimization problem in order to determine the optimal load/discharge schedule given energy prices.

NOMENCLATURE

BSS	Battery storage system
COE	Cost of energy (£)
FRA	Capital recovery factor (£)
C_(inv-i)	Capital cost of each component (£)
C_(O&M-i)	Operation and maintenance cost of each component (£)
C_(total-i)	Total investment cost (£)
D	Container diameter (m)
E_load	Load demand (W)
E_gen	Energy generated by renewable energies (W)
g	Gravitational acceleration (kg/s²)
G	Global irradiation (W/m²)
G_ref	Solar radiation under standard test conditions (W/m²)
H_p	Piston height (m)
H_c	Container height (m)
LCC	Life cycle cost
OCH	Discounted energy cost
LPSP	Likelihood of power loss

L_i	Life of each system (years)
NOCT	Nominal operating temperature of the cell (°C)
N_pv	Number of PVs
N_wt	Number of wind turbines
P_cal	Calorific value of the substrate (J/m^3)
P_(pv_rate)	Rated power (W)
P_(wind _rate)	Rated power (W)
r	Discount rate
SOC	State of charge
T_cref	Solar cell reference temperature (°C)
T_c	Nominal operating temperature (°C)
T_a	Ambient temperature (°C)
v	Wind speed (m/s)
v_cut in	Engagement speed (m/s)
v_(cut-out)	Cut speed (m/s)
v_rate	Nominal speed (m/s)
V_biogas	Biogas volume (m^3)
η_pv	Efficiency of photovoltaic module
β_t	Temperature coefficient of silicon cells
μ	System performance
ρ_rel	System density (kg/m^3)
ρ_piston	Piston density (kg/m^3)
ρ_water	Water density (kg/m^3)

REFERENCES

[1] S. Algarni, V. Tirth, T. Alqahtani, S. Alshehery, and P. Kshirsagar, "Contribution of renewable energy sources to the environmental impacts and economic benefits for sustainable development," *Sustainable Energy Technologies and Assessments*, vol. 56, p. 103098, Mar. 2023, doi: 10.1016/j.seta.2023.103098

[2] Irena. "Renewable energy integration in power grids," *Technology Brief*, 2015. www.irena.org/Publications

[3] S. O. Amrouche, D. Rekioua, T. Rekioua, and S. Bacha, "Overview of energy storage in renewable energy systems," *International Journal of Hydrogen Energy*, vol. 41, no. 45, pp. 20914–20927, 2016, doi: 10.1016/j.ijhydene.2016.06.243

[4] A. Emrani, A. Berrada, A. Arechkik, and M. Bakhouya, "Improved techno-economic optimization of an off-grid hybrid solar/wind/gravity energy storage system based on performance indicators," *Journal of Energy Storage*, vol. 49, p. 104163, 2022, doi: 10.1016/j.est.2022.104163

[5] A. Ameur, A. Berrada, and A. Emrani, "Dynamic forecasting model of a hybrid photovoltaic/gravity energy storage system for residential applications," *Energy and Buildings*, vol. 271, p. 112325, 2022, doi: 10.1016/j.enbuild.2022.112325

[6] P. T. Kapen, B. A. M. Nouadje, V. Chegnimonhan, G. Tchuen, and R. Tchinda, "Techno-economic feasibility of a PV/battery/fuel cell/electrolyzer/biogas hybrid system for energy and hydrogen production in the far north region of cameroon by using HOMER pro," *Energy Strategy Reviews*, vol. 44, p. 100988, 2022, doi: 10.1016/j.esr.2022.100988

[7]   A. Emrani, A. Berrada, and M. Bakhouya, "Optimal sizing and deployment of gravity energy storage system in hybrid PV-Wind power plant," *Renewable Energy*, vol. 183, pp. 12–27, Jan. 2022, doi: 10.1016/j.renene.2021.10.072

[8]   O. Schmidt, S. Melchior, A. Hawkes, and I. Staffell, "Projecting the future levelized cost of electricity storage technologies," *Joule*, vol. 3, no. 1, pp. 81–100, Jan. 2019, doi: 10.1016/j.joule.2018.12.008

[9]   A. Belderbos, E. Delarue, K. Kessels, and W. D'haeseleer, "Levelized cost of storage — Introducing novel metrics," *Energy Economics*, vol. 67, pp. 287–299, Sep. 2017, doi: 10.1016/j.eneco.2017.08.022

[10]   B. Battke, T. S. Schmidt, D. Grosspietsch, and V. H. Hoffmann, "A review and probabilistic model of lifecycle costs of stationary batteries in multiple applications," *Renewable and Sustainable Energy Reviews*, vol. 25, pp. 240–250, Sep. 2013, doi: 10.1016/j.rser.2013.04.023

[11]   M. Baumann, J. F. Peters, M. Weil, and A. Grunwald, "CO_2 footprint and life-cycle costs of electrochemical energy storage for stationary grid applications," *Energy Technol.*, vol. 5, no. 7, pp. 1071–1083, Jul. 2017, doi: 10.1002/ente.201600622

[12]   I. Staffell and M. Rustomji, "Maximising the value of electricity storage," *Journal of Energy Storage*, vol. 8, pp. 212–225, Nov. 2016, doi: 10.1016/j.est.2016.08.010

[13]   B. Zakeri and S. Syri, "Electrical energy storage systems: A comparative life cycle cost analysis," *Renewable and Sustainable Energy Reviews*, vol. 42, pp. 569–596, Feb. 2015, doi: 10.1016/j.rser.2014.10.011

[14]   P. Nikolaidis and A. Poullikkas, "Cost metrics of electrical energy storage technologies in potential power system operations," *Sustainable Energy Technologies and Assessments*, vol. 25, pp. 43–59, 2018, doi: 10.1016/j.seta.2017.12.001

[15]   H. Hesse, R. Martins, P. Musilek, M. Naumann, C. Truong, and A. Jossen, "Economic optimization of component sizing for residential battery storage systems," *Energies*, vol. 10, no. 7, p. 835, Jun. 2017, doi: 10.3390/en10070835

[16]   J. P. Fossati, A. Galarza, A. Martín-Villate, and L. Fontán, "A method for optimal sizing energy storage systems for microgrids," *Renewable Energy*, vol. 77, pp. 539–549, May 2015, doi: 10.1016/j.renene.2014.12.039

[17]   Y. Ru, J. Kleissl, and S. Martinez, "Storage size determination for grid-connected photovoltaic systems," *IEEE Transactions on Sustainable Energy*, vol. 4, no. 1, pp. 68–81, Jan. 2013, doi: 10.1109/TSTE.2012.2199339

[18]   M. Korpaas, A. T. Holen, and R. Hildrum, "Operation and sizing of energy storage for wind power plants in a market system," *International Journal of Electrical Power & Energy Systems*, vol. 25, no. 8, pp. 599–606, Oct. 2003, doi: 10.1016/S0142-0615(03)00016-4

[19]   C. Dusanter, "Energie solaire: définition et différents types d'exploitation," *Opéra Energie*, Oct. 19, 2020. https://opera-energie.com/energie-solaire/ (accessed Jun. 19, 2023).

[20]   "Énergie éolienne: fonctionnement, avantages, chiffres clés et enjeux," avr. – 12:00, 2011.   www.connaissancedesenergies.org/fiche-pedagogique/energie-eolienne (accessed Jun. 19, 2023).

[21]   T. Markvart, A. Fragaki, and J. N. Ross, "PV system sizing using observed time series of solar radiation," *Solar Energy*, vol. 80, no. 1, pp. 46–50, 2006, doi: https://doi.org/10.1016/j.solener.2005.08.011

5 Optimization and Operational Battery Management Systems for Residential Energy Systems

Truong Hoang Bao Huy, Daehee Kim,
Dieu Ngoc Vo, and Khoa Hoang Truong

5.1 INTRODUCTION

5.1.1 Overview of the Battery Management System

The battery management system (BMS) has emerged as a crucial component in the integration of energy storage systems, particularly in the context of residential applications. As renewable energy sources and electric vehicles gain prominence, homeowners seek sustainable energy [1, 2]. Battery energy storage systems (BESS) play a vital role in storing and utilizing generated power efficiently. The BMS serves as the intelligent control system that supervises the operation, performance, and safety of BESS. It monitors and manages parameters such as state of charge, voltage, temperature, and current, ensuring optimal charging and discharging cycles while protecting the battery from potential risks. Through its sophisticated algorithms and smart functionalities, the BMS maximizes energy utilization, extends battery lifespan, and enhances the overall performance of residential battery systems, empowering homeowners to embrace sustainable energy solutions with confidence.

Figure 5.1 describes a typical schematic of a smart home. It can be seen that the energy source of a smart home is supplied by the power grid and solar photovoltaic (PV) system [1]. The home loads include non-controllable and controllable appliances and plug-in hybrid electric vehicles (PEVs). These home loads are linked to energy sources via the important componence, that is, BESS. The BESS is intelligently managed to determine the optimal time for storing energy as reserves or releasing energy to the home loads and grid. During the operation of BESS, besides the state of charge, other parameters, such as voltage, temperature, and current, are also very important to monitor the overall performance of batteries. Accordingly, the battery system needs a management system to monitor and control the charging, discharging, and overall performance of batteries.

DOI: 10.1201/9781003441236-5

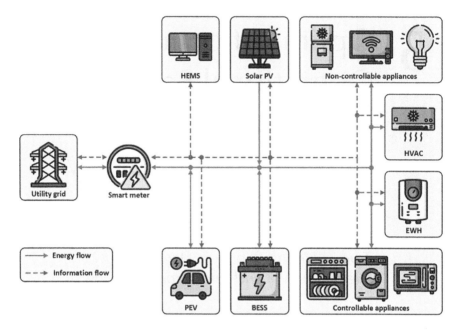

FIGURE 5.1 Typical schematic of a smart home.

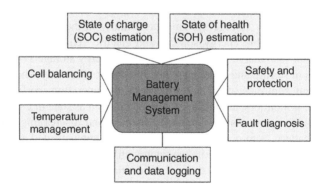

FIGURE 5.2 Functions of the battery management system.

The primary functions of a BMS are depicted in Figure 5.2. These functions are to ensure the safe and efficient operation of the battery and are described as follows.

1. *State of charge (SOC) estimation.* The BMS estimates the SOC of the battery. This parameter indicates the remaining capacity or energy available in the battery pack. This information is important for determining the battery's range, predicting its runtime, and preventing over-discharge or overcharge conditions.
2. *State of health (SOH) estimation.* The BMS assesses the overall health and degradation of the battery over time. It monitors parameters such as

temperature, voltage, and current to estimate the battery's remaining useful life and capacity.

3. *Cell balancing.* In multi-cell battery packs, individual cells can have different capacities or states of charge, leading to imbalances. The BMS ensures that each cell is charged and discharged uniformly, preventing overcharging of some cells and undercharging of others. This helps to maximize the overall capacity and lifespan of the battery pack.

4. *Temperature management.* Batteries can generate heat during charging and discharging, and excessive heat can degrade their performance and safety. The BMS monitors the temperature of the battery pack and implements measures like cooling or heating to maintain it within safe operating limits.

5. *Safety and protection.* The BMS incorporates safety measures to prevent over-charging, overdischarging, and over-temperature conditions. It may include features like short-circuit protection, overcurrent protection, and thermal management systems.

6. *Fault diagnosis.* The BMS continuously monitors the battery pack for any abnormalities or faults. In case of detected issues such as short circuits, overtemperature, or internal failures, the BMS can activate safety measures like disconnecting the battery or triggering alarms.

7. *Communication and data logging.* The BMS incorporates communication interfaces to exchange information with the main control system of the device or vehicle. They can also log data related to battery performance, including voltage, current, temperature, and other parameters, which is useful for diagnostics and analysis.

Overall, the BMS plays an important role in optimizing battery performance, extending battery life, ensuring safety, and maximizing the overall efficiency of battery-powered systems.

5.1.2 LITERATURE REVIEW

Among the functions of BMS, the most fundamental function is state estimation. Two of the most important indices employed in state estimates are state of charge (SOC) and state of health (SOH). This section reviews these indices.

SOC Estimation

The SOC parameter represents the remaining capacity that can be used in the current operational state of the battery. This parameter cannot be measured directly, and thus estimation of SOC is a challenging task in real applications [3]. As a result, numerous studies have been made to find reliable and secure approaches for SOC estimation during the past decade. In general, the approaches for SOC estimation can be classified into five methods, as shown in Figure 5.3.

The first method is the *look-up table method*. This method calculates the SOC value based on the mapping relationship between typical parameters (i.e., open-circuit voltage (OCV), impedance, internal resistance, etc.) and SOC [4]. For instance, the OCV look-up table method establishes the OCV-SOC table by measuring the OCV

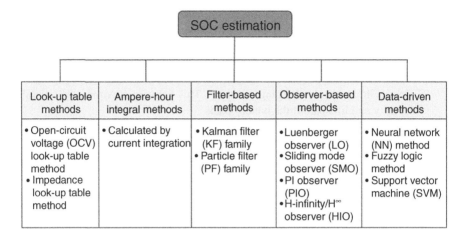

FIGURE 5.3 Method classification for SOC estimation.

under various SOC conditions, whereas in the impedance look-up table method, a non-linear fitting or parameter identification algorithm is used to identify several SOC-related parameters by delivering a specific frequency of current to the battery, and the impedance look-up table method is then established. In general, the look-up table method is straightforward and requires a non-linear relationship between SOC and OCV or impedance. Its drawbacks are low accuracy in practical applications due to sensor errors and an inability to satisfy real-time requirements.

The second method is the *ampere-hour integral method*. In this method, the battery's SOC is determined using the current integration [5]. The ampere-hour integral method is easy to use and requires low computational effort. However, the accuracy of this method mainly depends on the initial SOC and the sensor errors. As a result, using this method in combination with others, such as OCV look-up tables or model-based techniques, is a viable strategy for increasing the accuracy of SOC estimates [6].

The next method is the *filter-based method*. This approach can be classified into two different types: Kalman filter (KF) based methods and particle filter (PF) based methods. The KF-based methods are widely employed for SOC estimation with various variants such as linear Kalman filter (LKF), extended Kalman filter (EKF), sigma-point Kalman filter (SPKF), unscented Kalman filter (UKF), central difference Kalman filter (CDKF), and cubature Kalman filter (CKF). The LKF method is only appropriate for linear models. The EKF method is developed by approximating the linearization of the non-linear models to overcome the drawbacks of the LKF method. Thus, the accuracy of the EKF method depends on the error of approximation. As a result, the EKF method is inappropriate for high-order non-linear models. The SPKF and UKF methods are developed to avoid the complex Jacobian matrix and Gaussian noise computation; however, more computing capacity is required.

The KF method is constrained by the Gaussian distribution, and hence it is challenging to achieve a suitable filtering result for non-Gaussian distributions. To cope with non-Gaussian noise, a particle filter (PF) based approach is suggested and can provide high accuracy for battery state estimation. The PF method creates a set of discrete sampling points in the state space based on the empirical distribution of the system state vector and then modifies the particle's state and position in accordance with the measured values. The best particle state is then estimated by modifying the particle sets [7]. The unscented particle filter (UPF) method is an enhanced version of the PF. It employs UKF to enhance the sampling procedure of PF. In the UPF method, each particle's mean and variance are determined by the UKF using the function of approximate posterior filter density, and then the particle is resampled by the PF. The computation results embrace the latest observation information, and thus the PF sampling effect is enhanced [8]. Another improved version of PF is the cubature particle filter (CPF) method. In the CPF method, the mean and variance values of the non-linear random function are calculated directly using the volume method. A proposed density function is then created to produce weighted particles. Next, the minimum mean square error estimation of the system state is determined by calculating the mean value of the particles [9]. CPF uses the latest measurement information when generating particles, which improves the approximation degree of the posterior probability of the system state [10]. In general, the UPF- and CPF-based approaches are developed to enhance the particle sampling process and state probability approximation; however, the convergence rate is still a difficult problem.

In Figure 5.3, the third method for SOC estimation is the *observer-based method*. This method has a self-correction capability and provides high estimation accuracy with robustness [11]. One of the first observer-based methods is the Luenberger observer (LO) method [12]. This method can adapt to time-varying, non-linear, and linear models. In [13], an adaptive LO method was proposed for online SOC estimation, in which the observer gain was adaptively modified through a stochastic gradient approach. Another observer-based method is the sliding mode observer (SMO) method. This method is derived from the sliding mode controller (SMC) and provides strong tracking performance in the presence of model uncertainty and external interference. In SMO, a switching gain is created to maintain stability and convergence. A robust SMO was proposed in [14] for SOC estimation. In this study, the robust strategy incorporated a radial basis function neural network for learning the upper bound of the uncertain battery equivalent circuit model. The learned upper bound was used to adaptively adjust the switching gain of the robust SMO. The tuning process of the switching gain is to ensure the convergence of the SOC estimation errors to zero. Other improved versions based on the SMO method such as an adaptive gain SMO [15], a grey prediction-based fuzzy SMO [16], a super-twisting SMO [17], and so on have also been applied for SOC estimation. The next observer-based method that can be mentioned is the proportional-integral observer (PIO) method. This method is an efficient approach for estimating the state of systems with unknown input disturbances. In [18], the PIO method was proposed to estimate the SOC of lithium-ion batteries based on a simple-structure RC battery model. Another observer-based method is the H_∞ Observer (HIO) method. This method can maintain the robustness of the incorrect initial system state and unknown disturbance caused by incorrect or unknown statistical properties of modeling and measurement errors. In [19], the HIO method was suggested with dynamic gain for estimating the SOC of a pack of batteries, which can mitigate the negative effects of the non-Gaussian model and measurement errors. Several observer-based methods can also be mentioned for SOC estimation such as a recursive total least squares (RTLS) based observer [20], a discrete-time non-linear observer [21], and input-to-state stability (ISS) theory-based non-linear observer [22].

The last method mentioned in Figure 5.3 is the *data-driven method*. One of the data-driven methods commonly used for SOC estimation is the neural network (NN) method. In [23], the OCV-based method was proposed for SOC estimation by using the dual NN fusion battery models. The linear NN battery model was applied to determine the parameters of the first-order or second-order electrochemical models, while the back-propagation NN was used to capture the relationship between OCV and SOC. In [24], a load-classifying NN model was developed for estimating the SOC. With three neural networks trained in parallel, this method preliminary processes battery inputs and categorizes battery operation modes as idle, charge, and discharge. The authors in [25] estimated the SOC for lithium-ion batteries using long short-term memory (LSTM) recurrent neural network (RNN). In [26], a stacked LSTM network was suggested for SOC estimation based on the complex dynamics of batteries. Also, other improved versions of NN were applied for SOC estimation such as the deep feed-forward NN [26], the RNN with gated recurrent units [27], the multi-hidden-layer wavelet NN (WNN) [28], the fuzzy NN [29], and so on. The support vector machine (SVM) is one of the data-driven methods related to both classification and regression. In [30], an optimized SVM for regression was presented for SOC estimation. In [31], the least-square support vector machines (LSSVM) was combined with the adaptive unscented Kalman filters (AUKF) for estimating the battery's SOC.

SOH Estimation
The state of health (SOH) or remaining useful life (RUL) is an important metric for predicting lifetime and health, and it can indicate the current life status of batteries. The approaches for SOH estimation can be classified into three categories, as depicted in Figure 5.4.

The first category is the *measurement and analysis approach* which includes the direct measurement approach and the indirect analysis approach. The direct measurement approach directly measures critical parameters of the battery such as available capacity, internal resistance, and impedance. In [32], the coulomb counting method was used for SOH estimation by evaluating the maximum releasable capacity. In [33], the authors proposed an overview of methods based on impedance measurements to estimate the SOH. In [34], the electrochemical impedance spectroscopy (EIS) technique was employed to identify the online SOH by changing the impedance spectrum at a specific peak. Regarding the indirect analysis approach, in [35], a framework for battery SOH monitoring was proposed based on partial charging data. The incremental capacity analysis was used to identify the battery aging, and the support vector regression (SVR) algorithm was used to extract the robust aging signature. The simulation result showed that the SRV model based on data from a single cell can predict

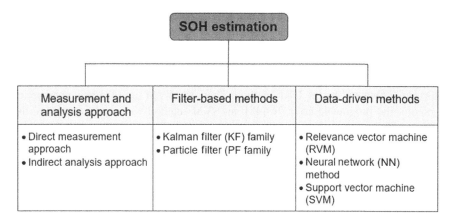

FIGURE 5.4   Method classification for SOH estimation.

the capacity degradation of seven other cells within a 1 percent error bound. The authors in [36] developed differential thermal voltammetry (DTV) to estimate the SOH. The DTV peak parameters that represent the specific chemical reaction can be used to investigate the aging process [37]. The differential mechanical parameter (DMP) analysis is an addition to the existing SOH estimation methods [38]. In [39], stress measurements were used to construct a DMP-based technique for estimating SOH. In general, the measurement and analysis approach is the most basic and straightforward method for calculating and analyzing the measured exterior parameters of batteries. Because this approach is open-loop, its accuracy is impacted by noise and environmental disturbance.

The second approach is *filter-based* methods including the Kalman filters (KF) family and the particle filters (PF) family. Regarding the KF family, in [40], the dual extended KF (DEKF) was applied to synchronously estimate both the SOC and SOH of the battery and model parameters. In [41], a hybrid algorithm that combines an adaptive unscented KF (AUKF) with a genetic algorithm (GA) associated with SVR (GA-SVR) was proposed to predict the remaining useful life (RUL) of the battery. The AUKF method was used to dynamically update the noise covariance while the SVR was applied to perform multistep prediction. The parameters of SVG were optimized by GA. Regarding the PF family, a gradient-correction PF was proposed to predict battery aging in [42]. The study in [43] introduced an unscented PF (UPF) to predict the battery remaining useful life (RUL) of lithium-ion batteries. The analysis results showed that the prediction achieved an error of less than 5 percent. In [44], the Gauss–Hermite PF method was proposed to estimate the RUL of the battery. To estimate the battery capacity, a state projection scheme was integrated. In [45], a RUL prediction for lithium-ion batteries was proposed using the exponential model and the PF. In [46], a prognostic method was developed to predict the RUL. To estimate capacity degradation, a state-space model of battery capacity was first built. Then, a spherical cubature PF (SCPF) was applied for solving the state-space model.

The last approach for battery SOH and RUL prediction is based on the *data-driven method*. In [47], a RUL prediction was proposed based on an error-correction scheme of a hybrid method. The solving method combined the algorithms including the UKF, complete ensemble empirical mode decomposition (CEEMD), and relevance vector machine (RVM). The CEEMD method provided a new error series based on a raw error series obtained by the UKF method. Then, the RVM used the new error series to predict the prognostic error. In [48], an SVM method was used to estimate the SOH and RUL. The authors in [49] introduced a SOH estimate approach based on the previous knowledge-based neural network (PKNN) and Markov chain to deal with uncertain external conditions and complex internal electrochemical processes. In [50], an online SOH estimation was carried out using a predictive diagnosis model by comparing partial charging curves with the stored SVMs. In [51], the Gaussian process regression (GPR) technique was proposed to forecast the battery SOH that outperformed existing data-driven methods. In general, the data-driven based approach is flexible with complicated models. However, its prediction results are susceptible to the training dataset.

5.1.3 PROSPECTS AND CHALLENGES

As energy storage technology becomes more widespread, there is a growing interest in enhancing advanced battery management technology. The battery is a complicated electrochemical system, posing challenges in system modeling and state estimation. This challenge significantly impacts the dependability and efficiency of the BMS. Thus, this section highlights some challenges and presents some research directions for the BMS.

Sensing Technology

The interior of the battery is an enclosed area, making it difficult to directly measure its various internal state and characteristics. Therefore, indirect approaches have been developed to estimate the battery state. However, these methods will have some errors in estimation or identification. To obtain more precise outcomes, advanced sensors are used to directly measure the internal states and parameters of the battery. For instance, fiber-optic sensors can be used to measure the internal states and temperature of the Li-ion battery pouch cells [52]. A buckled membrane sensor was presented to simultaneously monitor the characterization of stress and microstructure evolutions in the lithium-ion battery during its operation [53]. In [54], guided waves were used to precisely estimate the SOC and SOC of the lithium-ion battery using built-in piezoelectric sensors. In [55], high-precision contact-type displacement sensors are used to monitor the free swelling of the cell for SOC-imbalance detection. Future developments are expected to create advanced sensor technologies with characteristics of real-time, accuracy, and robustness. These advanced sensors in conjunction with multi-sensor information fusion technology can provide a deeper understanding of the internal structure and mechanism of the battery, thereby improving the BMS.

Thermal Management

Temperature has a significant impact on battery performance, safety, and lifespan. Predicting and analyzing thermal characteristics are very important to thermal management. There are few studies that focus on the electro-thermal model of lithium-ion batteries and the thermal abuse model of batteries [56]. Research techniques used to study the dispersion of current density within the battery electrodes and the issue of the battery thermal runaway produced by side reactions at high temperatures are still in the simulation phase. In general, there are still limitations to the analysis and research techniques for thermal failure and thermal abuse.

Fault Diagnosis

Fault diagnosis in BMS involves locating and examining any issues or anomalies that might arise inside the battery system. Through continuous monitoring of parameters like voltage, current, temperature, and SOC, the BMS employs artificial intelligence algorithms to detect deviations from normal operation. When anomalies are detected, algorithms analyze the data to determine the specific type of fault, its underlying cause, location, and potential implications. The BMS then generates alerts to notify operators, enabling swift actions to mitigate issues. The BMS records the fault data, analysis, and response actions for future reference and analysis. This information can aid in improving system design, maintenance strategies, and overall reliability. This process is crucial for maintaining the safety, efficiency, and reliability of battery systems, as well as preventing potential hazards.

Fault diagnosis in BMS suffers several challenges due to the diverse range of fault types such as short circuits, overcharging, undercharging, thermal runaway, and cell imbalances. Developing algorithms and methods that can effectively detect and differentiate between these diverse fault types is challenging. Noise data can cause difficulties in fault diagnosis. Monitoring data can be noisy and fluctuate due to environmental factors, sensor inaccuracies, and transient conditions. Another challenge is real-time processing. Fault diagnosis often requires real-time processing. As a result, efficient algorithms are needed to balance accuracy and speed. To overcome these challenges, a combination of advanced sensors technology, data analytics, machine learning, and system design needs to be developed to ensure effective and reliable fault diagnosis within BMS.

Communication

Communication is an essential aspect of BMS to ensure the smooth and coordinated operation of many components within a battery system. This involves the exchange of information and data between different parts of the system to monitor, control, and optimize battery performance and safety. On the one hand, communication is between individual battery cells or modules and the central BMS. Data of voltage, current, temperature, and SOC of each cell are continuously monitored and reported to the BMS for real-time assessment and control. These data help identify any deviations from normal behavior, enabling quick response to prevent issues such as overcharging, overheating, or cell imbalances. On the other hand, communication within the BMS also extends to external interfaces, allowing the system to interact with external devices, such as chargers, inverters, or energy management systems. This facilitates coordinated charging, discharging, optimizing the overall efficiency and lifespan of the battery system.

Nowadays, the emergence of vehicle-to-grid (V2G) technology along with the rise of the high-speed 5G information era poses new requirements for BMS. The BMS scheme based on big data and cloud computing platforms will be a future research direction.

5.2 HOME ENERGY MANAGEMENT SYSTEM

This section describes a basic HEMS, in which the operations of BESS is optimized. HEMS includes PV integration as a reserve power supply, BESS, and home appliances in a typical smart home, as shown in Figure 5.1. HEMS is employed to effectively regulate energy consumption and ensure a consistent power supply for the energy demands of homeowners.

In HEMS, the communication network plays a vital role in facilitating the exchange of information among the HEMS itself, the utility grid, and the user. The HEMS is responsible for receiving, regulating, and managing data and communications from users, devices, and external channels. The data fed into the HEMS include forecast electricity prices, predicted solar irradiation, predicted energy consumption of home appliances, and the current SOC of the BESS. Based on the input data, the optimization algorithm integrated into the HEMS analyzes the gathered information and optimizes the operational schedule of the BESS. Subsequently, the HEMS transmits control signals to the BESS.

5.2.1 SYSTEM MODELING

Grid

In a prosumer environment, a HEMS has the option to buy power from or sell power to the utility grid. The concept of grid modeling [57, 58] is presented in Equations (5.1–5.2)

$$0 \le P_t^{G2H} \le u_t^{G2H} \cdot \overline{P}^{G2H}; \quad \forall t = 1,2,...,T \tag{5.1}$$

$$0 \le P_t^{H2G} \le u_t^{H2G} \cdot \overline{P}^{H2G}; \quad \forall t = 1,2,...,T \tag{5.2}$$

where P_t^{G2H} is the power purchased from the grid at time slot t; P_t^{H2G} is the power sold back to the grid at time slot t; $\overline{P}^{G2H}$ and $\overline{P}^{H2G}$ are the maximum powers that can be exchanged between HEMS and the utility grid, respectively; and u_t^{G2H} and u_t^{H2G} are the binary variables for the purchase and sell modes of the HEMS with the utility grid at time slot t, respectively.

The constraint in Equation (5.3) describes a process that is impossible for the buying and selling of energy to occur at the same time [57, 58].

$$0 \le u_t^{G2H} + u_t^{H2G} \le 1; \quad \forall t = 1,2,...,T \tag{5.3}$$

Solar PV System

The HEMS can determine the anticipated solar PV output for the day ahead based on daily weather data of solar irradiance. Therefore, the maximum power produced from the solar PV at time slot t is calculated in Equation (5.4) [58, 59].

$$\phi_t^{PV} = \upsilon_t \cdot \overline{P}^{PV} \cdot \eta^{PV} \cdot \Delta t; \quad \forall t = 1,2,\ldots,T \tag{5.4}$$

where $\overline{P}^{PV}$ is the solar PV peak power, η^{PV} is the conversion efficiency, and υ_t is the solar irradiance at time slot t.

BESS

A BESS has two modes of operation including charging or discharging. When the BESS is in discharge mode, it is treated as a load; in contrast, BESS is considered as the source for charging mode. Hence, the SOC of the BESS is modeled as shown Equation (5.5) [60, 61].

$$\varepsilon_t^{BESS} = \varepsilon_{t-1}^{BESS} + \left(\eta^{BESS} \cdot P_t^{BESS,ch} - \frac{P_t^{BESS,dch}}{\eta^{BESS}} \right) \cdot \Delta t; \quad \forall t = 1,2,\ldots,T \tag{5.5}$$

where ε_t^{BESS} and ε_{t-1}^{BESS} are the SOC of the BESS at time slot t and time slot $(t-1)$, respectively; $P_t^{BESS,ch}$ is the energy charged into the BESS at time slot t; $P_t^{BESS,dch}$ is the energy discharged from the BESS at time slot t; and η^{BESS} is the charge/discharge efficiency of the BESS.

The SOC of the BESS is limited by its maximum capacity and depth of discharge (DOD) as follows:

$$(1 - DOD^{BESS}) \cdot \overline{\varepsilon}^{BESS} \leq \varepsilon_t^{BESS} \leq \overline{\varepsilon}^{BESS}; \quad \forall t = 1,2,\ldots,T \tag{5.6}$$

where $\overline{\varepsilon}^{BESS}$ is the maximum capacity of the BESS, and DOD^{BESS} is the DOD of the BESS.

The powers for charging and discharging must be limited to rated power as follows [57, 59]:

$$0 \leq P_t^{BESS,ch} \leq u_t^{BESS,ch} \cdot \overline{P}^{BESS,ch}; \quad \forall t = 1,2,\ldots,T \tag{5.7}$$

$$0 \leq P_t^{BESS,dch} \leq u_t^{BESS,dch} \cdot \overline{P}^{BESS,dch}; \quad \forall t = 1,2,\ldots,T \tag{5.8}$$

where $\overline{P}^{BESS,ch}$ and $\overline{P}^{BESS,dch}$ are the rated charge and discharge powers of the BESS, respectively.

The charge and discharge modes of BESS must not occur at the same time, which can be expressed as follows [57, 59]:

$$0 < u_t^{BESS,ch} + u_t^{BESS,dch} \leq 1; \quad \forall t = 1,2,\ldots,T \tag{5.9}$$

where $u_t^{BESS,ch}$ and $u_t^{BESS,dch}$ are binary variables for the BESS charge and discharge modes at time slot t, respectively.

It is assumed that the SOC of the BESS should be at its peak capacity at the beginning and end of the daily planning cycle, which can be given as follows [59]:

$$\varepsilon_1^{BESS} = \varepsilon_T^{BESS} = \overline{\varepsilon}^{BESS} \tag{5.10}$$

Energy Balance

In the proposed HEMS model, the energy balance is expressed as follows [62]:

$$P_t^{G2H} + P_t^{PV} + P_t^{BESS,dch} = P_t^{H2G} + P_t^{BESS,ch} + P_t^{load}; \quad \forall t = 1, 2, \ldots, T \tag{5.11}$$

where P_t^{load} is the electricity consumption of household appliances at time slot t.

5.2.2 OBJECTIVE FRAMEWORK

The proposed HEMS aims to minimize the daily energy costs by scheduling the charge and discharge operations of the BESS. Therefore, the proposed HEMS model is defined as a MILP optimization problem as follows:

$$Objective\ function: \quad \min_x \left(\sum_{t=1}^{T} \left\{ \Delta t \cdot \left(\lambda_t^{G2H} \cdot P_t^{G2H} - \lambda_t^{H2G} \cdot P_t^{H2G} \right) \right\} \right) \tag{5.12}$$

Inequality constraint: Equations (5.1), (5.2), (5.3), (5.6), (5.7), (5.8), (5.9)

Equality constraint: Equations (5.4), (5.5), (5.10), (5.11)

$$Decision\ variables: \ x = \left\{ \begin{matrix} P_t^{G2H}, P_t^{H2G}, P_t^{BESS,ch}, P_t^{BESS,dch}, \\ u_t^{G2H}, u_t^{H2G}, u_t^{BESS,ch}, u_t^{BESS,dch}, \end{matrix} \right\}; \quad \forall t = 1, 2, \ldots, T \tag{5.13}$$

where λ_t^{G2H} is the purchase price of electricity from the grid at time slot t, and λ_t^{H2G} is the selling price of electricity to the grid at time slot t.

5.3 SIMULATION RESULTS

The simulation results of the proposed HEMS modeling are presented for a typical home with a solar PV panel and a BESS (as shown in Figure 5.1). Table 5.1 includes the data for the grid, solar PV panel, and BESS. Day-ahead data for electricity price, solar irradiation, and energy consumption are considered in this study. Figure 5.5 depicts the hourly electricity price extracted from Commonwealth Edison Company on July 18, 2016 [63]. Assume that the selling price is 0.9 times the purchase price. Figure 5.6 displays solar irradiance on July 18, 2016, in Madrid (Spain), taken from European Commission [64]. Hourly energy consumption of a typical smart home is taken from open source in [65], as shown in Figure 5.7. The simulations cover a period of 24 hours, divided into 48 time slots with a time step of 30 minutes. The MILP model of the HEMS is implemented in Python and solved using the Gurobi optimizer.

TABLE 5.1
Data for the grid, solar PV panel, and BESS

Item	Parameter	Value
Grid	$\bar{P}^{G2H}/\bar{P}^{H2G}$	10 /4 kW
Solar PV panel	$\bar{P}^{PV}$	1 kW
	η^{PV}	0.167
BESS	$\bar{\varepsilon}^{BESS}$	5 kWh
	$\bar{P}^{BESS,ch}/\bar{P}^{BESS,dch}$	2 kW / 2 kW
	η^{BESS}	0.98
	DOD^{BESS}	0.70

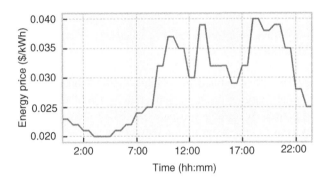

FIGURE 5.5   Hourly electricity price.

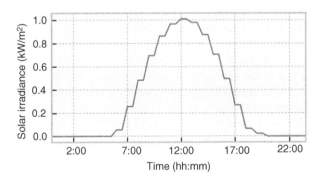

FIGURE 5.6   Hourly solar irradiance.

This study takes into account the following assumptions:

- The impacts of BESS charging and discharging cycles on battery degradation are neglected [66];
- The maintenance expenses of the BESS and solar PV panels are disregarded [62, 66]; and

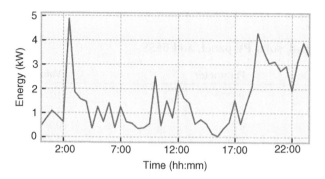

FIGURE 5.7 Hourly energy consumption.

- All necessary forecast information (i.e., electricity price and solar irradiation) is provided in full [57].

The proposed HEMS is applied to schedule optimal operations of the BESS. Figure 5.8 depicts the charging/discharging operations and SOC of the BESS over 24-hour scheduling period. According to Figure 5.8, BESS stores energy from the grid when the energy prices are low during the initial time slots of the scheduling cycle. BESS charges continuously until it reaches its maximum capacity before the solar PV panel starts generating energy in the first hour. At these time slots, electricity demand is relatively low. Subsequently, BESS releases stored energy during time slots of high energy prices and high electricity demand. Accordingly, BESS deeply discharges to deliver the permitted energy to the grid from 10:00 in accordance with the increasing trend of the energy price. It is clear that the SOC of the BESS is always guaranteed to not exceed its DOD and maximum capacity. Hence, overcharging/overdischarging of BESS is avoided. The SOC of the BESS is also kept to its minimum at the initial and final time slots. This shows the accuracy of the proposed HEMS.

Figure 5.9 depicts the energy exchanged between the home and grid. Total energy purchased increases slightly from 28.0468 kW (base case) to 30.8075 kW. However, total energy sold also increases significantly from 5.7062 kW (base case) to 8.0485 kW. Although there is not much change in the net energy exchanged with the grid, BESS integration significantly reduces household electricity cost. Indeed, the electricity cost is greatly reduced from $0.6610 to $0.5683, corresponding to a 14.02 percent reduction compared to the base case without BESS. Without BESS, surplus energy from solar PV is forced to be sold to the grid at any energy price to avoid wasting energy. With the storage capacity of BESS, this surplus energy can be stored in BESS to supply the load or sell to the grid in the following hours. Moreover, residential users can participate more actively in the market by purchasing electricity during off-peak times to fully charge BESS and selling energy to the grid during peak times to generate profits. Therefore, the proposed HEMS effectively optimizes the charging and discharging operations of BESS, contributing to lower electricity bills.

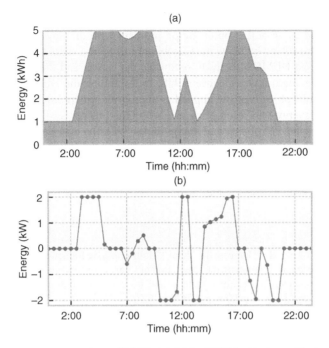

FIGURE 5.8 Optimal operations of BESS: (a) SOC of BESS; (b) charge/discharge power of BESS (negative energy means BESS is discharging).

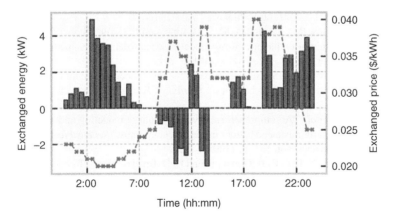

FIGURE 5.9 Exchanged energy between the home and grid.

5.4 CONCLUSION

This chapter has presented the advanced battery management system, including its concepts and cutting-edge strategies to optimize energy management in residential settings. In this chapter, a basic HEMS model has been proposed, wherein solar PV panels and BESS were fully utilized in a typical smart home. The proposed HEMS was formulated as an MILP optimization problem with the objective of minimum energy costs and solved by using the Gurobi optimizer. The simulation result showed that the proposed HEMS effectively optimizes the charge and discharge operations of BESS, achieving a lower electricity bill. In the future, the HEMS model can be extended to multi-objectives and applied to microgrids, where multiple smart homes can exchange energy with each other and with other energy systems, such as wind power plants, solar PV plants, and energy hubs. With the development of big data, cloud computing, the Internet of Things (IoT), and other emerging techniques, advanced machine learning algorithms are suitable to embed into the advanced BMS to fully unlock the potential of residential battery systems, contributing to a greener and more sustainable energy landscape.

REFERENCES

[1] T. H. B. Huy, H. T. Dinh, and D. Kim, "Multi-objective framework for a home energy management system with the integration of solar energy and an electric vehicle using an augmented ε-constraint method and lexicographic optimization," *Sustainable Cities and Society*, vol. 88, p. 104289, 2023.

[2] T. H. B. Huy, H. Truong Dinh, D. Ngoc Vo, and D. Kim, "Real-time energy scheduling for home energy management systems with an energy storage system and electric vehicle based on a supervised-learning-based strategy," *Energy Conversion and Management*, vol. 292, p. 117340, 2023.

[3] Z. Ren, C. Du, Z. Wu, J. Shao, and W. Deng, "A comparative study of the influence of different open circuit voltage tests on model-based state of charge estimation for lithium-ion batteries," *International Journal of Energy Research*, vol. 45, no. 9, pp. 13692–13711, 2021. https://doi.org/10.1002/er.6700

[4] D. N. T. How, M. A. Hannan, M. S. H. Lipu, and P. J. Ker, "State of charge estimation for lithium-ion batteries using model-based and data-driven methods: A review," *IEEE Access*, vol. 7, pp. 136116–136136, 2019.

[5] Y. Zhang, W. Song, S. Lin, and Z. Feng, "A novel model of the initial state of charge estimation for LiFePO4 batteries," *Journal of Power Sources*, vol. 248, pp. 1028–1033, 2014.

[6] Y. Wang *et al.*, "A comprehensive review of battery modeling and state estimation approaches for advanced battery management systems," *Renewable and Sustainable Energy Reviews*, vol. 131, p. 110015, 2020.

[7] A. Tulsyan, Y. Tsai, R. B. Gopaluni, and R. D. Braatz, "State-of-charge estimation in lithium-ion batteries: A particle filter approach," *Journal of Power Sources*, vol. 331, pp. 208–223, 2016.

[8] R. Van Der Merwe, A. Doucet, N. De Freitas, and E. Wan, "The unscented particle filter," Technical Report CUED/F-INFENG/TR 380, Cambridge University Engineering Department, 2000.

[9] R. Guo, Q. Gan, J. Zhang, K. Guo, and J. Dong, "Huber cubature particle filter and online state estimation," *Proceedings of the Institution of Mechanical Engineers, Part I: Journal of Systems and Control Engineering*, vol. 231, no. 3, pp. 158–167, 2017.

[10] R. Guo, Q. Gan, J. Zhang, K. Guo, and J. Dong, "Huber cubature particle filter and online state estimation," *Proceedings of the Institution of Mechanical Engineers, Part I: Journal of Systems and Control Engineering*, vol. 231, no. 3, pp. 158–167, 2017.

[11]   P. Shrivastava, T. K. Soon, M. Y. I. B. Idris, and S. Mekhilef, "Overview of model-based online state-of-charge estimation using Kalman filter family for lithium-ion batteries," *Renewable and Sustainable Energy Reviews*, vol. 113, p. 109233, 2019.

[12]   D. Luenberger, "Observers for multivariable systems," *IEEE Transactions on Automatic Control*, vol. 11, no. 2, pp. 190–197, 1966.

[13]   X. Hu, F. Sun, and Y. Zou, "Estimation of state of charge of a lithium-ion battery pack for electric vehicles using an adaptive Luenberger observer," *Energies*, vol. 3, no. 9, pp. 1586–1603, 2010.

[14]   X. Chen, W. Shen, M. Dai, Z. Cao, J. Jin, and A. Kapoor, "Robust adaptive sliding-mode observer using RBF neural network for lithium-ion battery state of charge estimation in electric vehicles," *IEEE Transactions on Vehicular Technology*, vol. 65, no. 4, pp. 1936–1947, 2016.

[15]   X. Chen, W. Shen, Z. Cao, and A. Kapoor, "A novel approach for state of charge esti-mation based on adaptive switching gain sliding mode observer in electric vehicles," *Journal of Power Sources*, vol. 246, pp. 667–678, 2014.

[16]   D. Kim, T. Goh, M. Park, and S. W. Kim, "Fuzzy sliding mode observer with grey prediction for the estimation of the state-of-charge of a lithium-ion battery," *Energies*, vol. 8, no. 11, pp. 12409–12428, 2015.

[17]   Y. Huangfu, J. Xu, D. Zhao, Y. Liu, and F. Gao, "A novel battery state of charge esti-mation method based on a super-twisting sliding mode observer," *Energies*, vol. 11, no. 5, p. 1211, 2018.

[18]   J. Xu, C. C. Mi, B. Cao, J. Deng, Z. Chen, and S. Li, "The state of charge esti-mation of lithium-ion batteries based on a proportional-integral observer," *IEEE Transactions on Vehicular Technology*, vol. 63, no. 4, pp. 1614–1621, 2014.

[19]   Q. Zhu, L. Li, X. Hu, N. Xiong, and G. D. Hu, "H_{∞}-based nonlinear observer design for state of charge estimation of lithium-ion battery with polyno-mial parameters," *IEEE Transactions on Vehicular Technology*, vol. 66, no. 12, pp. 10853–10865, 2017.

[20]   Z. Wei, C. Zou, F. Leng, B. H. Soong, and K.-J. Tseng, "Online model identifica-tion and state-of-charge estimate for lithium-ion battery with a recursive total least squares-based observer," *IEEE Transactions on Industrial Electronics*, vol. 65, no. 2, pp. 1336–1346, 2017.

[21]   W. Li, L. Liang, W. Liu, and X. Wu, "State of charge estimation of lithium-ion batteries using a discrete-time nonlinear observer," *IEEE Transactions on Industrial Electronics*, vol. 64, no. 11, pp. 8557–8565, 2017.

[22]   Y. Ma, B. Li, G. Li, J. Zhang, and H. Chen, "A nonlinear observer approach of SOC estimation based on hysteresis model for lithium-ion battery," *IEEE/CAA Journal of Automatica Sinica*, vol. 4, no. 2, pp. 195–204, 2017.

[23]   X. Dang, L. Yan, K. Xu, X. Wu, H. Jiang, and H. Sun, "Open-circuit voltage-based state of charge estimation of lithium-ion battery using dual neural network fusion battery model," *Electrochimica Acta*, vol. 188, pp. 356–366, 2016.

[24]   S. Tong, J. H. Lacap, and J. W. Park, "Battery state of charge estimation using a load-classifying neural network," *Journal of Energy Storage*, vol. 7, pp. 236–243, 2016.

[25]   E. Chemali, P. J. Kollmeyer, M. Preindl, R. Ahmed, and A. Emadi, "Long short-term memory networks for accurate state-of-charge estimation of Li-ion batteries," *IEEE Transactions on Industrial Electronics*, vol. 65, no. 8, pp. 6730–6739, 2017.

[26]   F. Yang, X. Song, F. Xu, and K. L. Tsui, "State-of-charge estimation of lithium-ion batteries via long short-term memory network," *IEEE Access*, vol. 7, pp. 53792–53799, 2019.

[27] F. Yang, W. Li, C. Li, and Q. Miao, "State-of-charge estimation of lithium-ion batteries based on gated recurrent neural network," *Energy*, vol. 175, pp. 66–75, 2019.

[28] B. Xia *et al.*, "State of charge estimation of lithium-ion batteries using optimized Levenberg-Marquardt wavelet neural network," *Energy*, vol. 153, pp. 694–705, 2018.

[29] C. Burgos, D. Sáez, M. E. Orchard, and R. Cárdenas, "Fuzzy modelling for the state-of-charge estimation of lead-acid batteries," *Journal of Power Sources*, vol. 274, pp. 355–366, 2015.

[30] J. N. Hu *et al.*, "State-of-charge estimation for battery management system using optimized support vector machine for regression," *Journal of Power Sources*, vol. 269, pp. 682–693, 2014.

[31] J. Meng, G. Luo, and F. Gao, "Lithium polymer battery state-of-charge estimation based on adaptive unscented Kalman filter and support vector machine," *IEEE Transactions on Power Electronics*, vol. 31, no. 3, pp. 2226–2238, 2016.

[32] K. S. Ng, C.-S. Moo, Y.-P. Chen, and Y.-C. Hsieh, "Enhanced coulomb counting method for estimating state-of-charge and state-of-health of lithium-ion batteries," *Applied Energy*, vol. 86, no. 9, pp. 1506–1511, 2009.

[33] H. Blanke *et al.*, "Impedance measurements on lead–acid batteries for state-of-charge, state-of-health and cranking capability prognosis in electric and hybrid electric vehicles," *Journal of Power Sources*, vol. 144, no. 2, pp. 418–425, 2005.

[34] M. Galeotti, C. Giammanco, L. Cinà, S. Cordiner, and A. D. Carlo, "Diagnostic methods for the evaluation of the state of health (SOH) of NiMH batteries through electrochemical impedance spectroscopy," in *2014 IEEE 23rd International Symposium on Industrial Electronics (ISIE)*, 2014, pp. 1641–1646.

[35] C. Weng, Y. Cui, J. Sun, and H. Peng, "On-board state of health monitoring of lithium-ion batteries using incremental capacity analysis with support vector regression," *Journal of Power Sources*, vol. 235, pp. 36–44, 2013.

[36] Y. Merla, B. Wu, V. Yufit, N. P. Brandon, R. F. Martinez-Botas, and G. J. Offer, "Novel application of differential thermal voltammetry as an in-depth state-of-health diagnosis method for lithium-ion batteries," *Journal of Power Sources*, vol. 307, pp. 308–319, 2016.

[37] B. Wu, V. Yufit, Y. Merla, R. F. Martinez-Botas, N. P. Brandon, and G. J. Offer, "Differential thermal voltammetry for tracking of degradation in lithium-ion batteries," *Journal of Power Sources*, vol. 273, pp. 495–501, 2015.

[38] L. W. Sommer *et al.*, "Monitoring of intercalation stages in lithium-ion cells over charge-discharge cycles with fiber optic sensors," *Journal of The Electrochemical Society*, vol. 162, no. 14, p. A2664, 2015.

[39] J. Cannarella and C. B. Arnold, "State of health and charge measurements in lithium-ion batteries using mechanical stress," *Journal of Power Sources*, vol. 269, pp. 7–14, 2014.

[40] N. Wassiliadis *et al.*, "Revisiting the dual extended Kalman filter for battery state-of-charge and state-of-health estimation: A use-case life cycle analysis," *Journal of Energy Storage*, vol. 19, pp. 73–87, 2018.

[41] Z. Xue, Y. Zhang, C. Cheng, and G. Ma, "Remaining useful life prediction of lithium-ion batteries with adaptive unscented kalman filter and optimized support vector regression," *Neurocomputing*, vol. 376, pp. 95–102, 2020.

[42] X. Tang, K. Liu, X. Wang, B. Liu, F. Gao, and W. D. Widanage, "Real-time aging trajectory prediction using a base model-oriented gradient-correction particle filter for Lithium-ion batteries," *Journal of Power Sources*, vol. 440, p. 227118, 2019.

[43] Q. Miao, L. Xie, H. Cui, W. Liang, and M. Pecht, "Remaining useful life prediction of lithium-ion battery with unscented particle filter technique," *Microelectronics Reliability*, vol. 53, no. 6, pp. 805–810, 2013.

[44] C. Hu, G. Jain, P. Tamirisa, and T. Gorka, "Method for estimating capacity and predicting remaining useful life of lithium-ion battery," *Applied Energy*, vol. 126, pp. 182–189, 2014.

[45] L. Zhang, Z. Mu, and C. Sun, "Remaining useful life prediction for lithium-ion batteries based on exponential model and particle filter," *IEEE Access*, vol. 6, pp. 17729–17740, 2018.

[46] D. Wang, F. Yang, K. L. Tsui, Q. Zhou, and S. J. Bae, "Remaining useful life prediction of lithium-ion batteries based on spherical cubature particle filter," *IEEE Transactions on Instrumentation and Measurement*, vol. 65, no. 6, pp. 1282–1291, 2016.

[47] Y. Chang, H. Fang, and Y. Zhang, "A new hybrid method for the prediction of the remaining useful life of a lithium-ion battery," *Applied Energy*, vol. 206, pp. 1564–1578, 2017.

[48] A. Nuhic, T. Terzimehic, T. Soczka-Guth, M. Buchholz, and K. Dietmayer, "Health diagnosis and remaining useful life prognostics of lithium-ion batteries using data-driven methods," *Journal of Power Sources*, vol. 239, pp. 680–688, 2013.

[49] H. Dai, G. Zhao, M. Lin, J. Wu, and G. Zheng, "A novel estimation method for the state of health of lithium-ion battery using prior knowledge-based neural network and Markov chain," *IEEE Transactions on Industrial Electronics*, vol. 66, no. 10, pp. 7706–7716, 2019.

[50] X. Feng *et al.*, "Online state-of-health estimation for Li-Ion battery using partial charging segment based on support vector machine," *IEEE Transactions on Vehicular Technology*, vol. 68, no. 9, pp. 8583–8592, 2019.

[51] R. R. Richardson, M. A. Osborne, and D. A. Howey, "Gaussian process regression for forecasting battery state of health," *Journal of Power Sources*, vol. 357, pp. 209–219, 2017.

[52] A. Ganguli *et al.*, "Embedded fiber-optic sensing for accurate internal monitoring of cell state in advanced battery management systems part 2: Internal cell signals and utility for state estimation," *Journal of Power Sources*, vol. 341, pp. 474–482, 2017.

[53] H. Jung, C. F. Lin, K. Gerasopoulos, G. Rubloff, and R. Ghodssi, "A buckled membrane sensor for in situ mechanical and microstructure analysis of li-ion battery electrodes," in *2015 Transducers – 2015 18th International Conference on Solid-State Sensors, Actuators and Microsystems (TRANSDUCERS)*, 2015, pp. 1953–1956.

[54] P. Ladpli, F. Kopsaftopoulos, and F.-K. Chang, "Estimating state of charge and health of lithium-ion batteries with guided waves using built-in piezoelectric sensors/actuators," *Journal of Power Sources*, vol. 384, pp. 342–354, 2018.

[55] Y. Kim, N. A. Samad, K. Y. Oh, J. B. Siegel, B. I. Epureanu, and A. G. Stefanopoulou, "Estimating state-of-charge imbalance of batteries using force measurements," in *2016 American Control Conference (ACC)*, 2016, pp. 1500–1505.

[56] Y. Hu, S. Yurkovich, Y. Guezennec, and B. J. Yurkovich, "Electro-thermal battery model identification for automotive applications," *Journal of Power Sources*, vol. 196, no. 1, pp. 449–457, 2011.

[57] M. Tostado-Véliz, P. Arévalo, S. Kamel, H. M. Zawbaa, and F. Jurado, "Home energy management system considering effective demand response strategies and uncertainties," *Energy Reports*, vol. 8, pp. 5256–5271, 2022.

[58]   M. Tostado-Véliz, D. Icaza-Alvarez, and F. Jurado, "A novel methodology for optimal sizing photovoltaic-battery systems in smart homes considering grid outages and demand response," *Renewable Energy*, vol. 170, pp. 884–896, 2021.

[59]   M. Tostado-Véliz, S. Gurung, and F. Jurado, "Efficient solution of many-objective Home Energy Management systems," *International Journal of Electrical Power & Energy Systems*, vol. 136, p. 107666, 2022.

[60]   I. Alsaidan, A. Khodaei, and W. Gao, "A comprehensive battery energy storage optimal sizing model for microgrid applications," *IEEE Transactions on Power Systems*, vol. 33, no. 4, pp. 3968–3980, 2018.

[61]   P. Arévalo, M. Tostado-Véliz, and F. Jurado, "A novel methodology for comprehensive planning of battery storage systems," *Journal of Energy Storage*, vol. 37, p. 102456, 2021.

[62]   M. Shafie-Khah and P. Siano, "A stochastic home energy management system considering satisfaction cost and response fatigue," *IEEE Transactions on Industrial Informatics*, vol. 14, no. 2, pp. 629–638, 2018.

[63]   *ComEd's Hourly Pricing Program*. Retrieved September 20, 2023. Available: https://hourlypricing.comed.com/

[64]   *JRC Photovoltaic Geographical Information System (PVGIS)—European Commission*. Retrieved September 20, 2023. Available: https://re.jrc.ec.europa.eu/pvg_tools/en/tools.html

[65]   Singh T. *Smart Home Dataset with Weather Information*. Available: www.kaggle.com/taranvee/smart-home-dataset-with-weather-information

[66]   N. G. Paterakis, O. Erdinç, A. G. Bakirtzis, and J. P. S. Catalão, "Optimal household appliances scheduling under day-ahead pricing and load-shaping demand response strategies," *IEEE Transactions on Industrial Informatics*, vol. 11, no. 6, pp. 1509–1519, 2015.

6 Regulatory Settings and Policy Options for Battery-Integrated Residential Systems

Swasti Swadha, Bhavna Jangid, Rohini Haridas,
Shubham Yadav, Chandra Prakash,
Ajay Verma, Arun Nayak, Shivani Garg,
Parul Mathuria, and Rohit Bhakar

6.1 INTRODUCTION

Residential consumers constitute a significant part of the distribution network [1]. They need power to use different electrical appliances and charge their electric vehicles (EVs). The concern for them is to get uninterrupted supplies at affordable costs. Their dependency on power supplies is associated with the adoption of battery storage. At the residential level, static battery storage is used to provide power in residences, either connected to a grid supply or with rooftop solar photovoltaic (PV). In addition, dynamic battery storage is used in EVs.

Battery storage has been used for a variety of reasons, not least ensuring power supply at different times of the day by charging/discharging during the off-peak and peak hours, respectively. Secondly, it helps reduce the dependency of residential consumers on the grid in countries where the functioning of power utilities is unsatisfactory [2]. In this regard, an increasing percentage of prosumers and their realization of the benefits of rooftop solar integrated with battery storage will help to save on electricity bills [3]. The use of battery storage accelerates the rooftop solar PV self-consumption growth rate. It stores excess generation, avoiding possible curtailments. Stored energy can be used during cloudy days when there is little or no solar irradiance. Storage of the excess PV generation rather than exporting it to the grid benefits the utilities through a reduction in distribution grid issues as well as the need for network reinforcements.

The third benefit is that storage adds value to our environment by promoting large-scale integration of rooftop solar. As an effect, this has increased significantly worldwide, with an annual average growth rate of about 50 percent from 2010 to 2020, and it continues to grow fast [4]. Similarly, batteries used in EVs also help in decarbonization by replacing the use of diesel and petrol. Lastly, the use of battery storage in home energy management systems (HEMS) or grid interactive efficient buildings

(GEBs) will be helpful in managing power consumption at the residential level. These are buildings equipped with various energy-efficient equipment as well as a bidirectional communication interface. This intelligently controlled system facilitates demand response (DR) to provide services to the grid when required. Battery storage can be instrumental in enhancing this process, as stored power provides flexibility for residential consumers to easily shift their consumption patterns from peak hours consistent with the grid requirements.

Few programs have been introduced to control the pattern of electricity usage and energy conservation at the residential level. However, there is a lack of policies that can ensure the active utilization of battery storage at a voluntary level by residential consumers to fulfill DR obligations, benefiting the utilities. Similarly, policies aimed at promoting increased integration of rooftop solar PV have been introduced in an attempt to reduce dependency on fossil fuels. But this has gradually led to congestion issues at the distribution level. There is a need for a mechanism that can seek to solve this concern through effective control and utilization of battery storage.

The chapter recommends policies and regulations for bringing together residential battery storage systems: static and dynamic, through aggregation. This will aim to benefit the major stakeholders of aggregated battery systems, that is, the residential consumers, power utilities, and the environment, in an efficient way. The goal can be achieved by emphasizing this aggregation using information and communication technologies (ICT) for bidirectional control. The high cost of battery storage is still a significant hindrance in this process [5]. Hence, there is also a need for market policies to reduce the cost and quality of battery storage.

The chapter is structured as follows. Section 6.2 enumerates various stakeholders in the aggregated network. Section 6.3 discusses the benefits of battery storage aggregation for promoting integration, and Section 6.4 mentions the challenges of aggregation. Section 6.5 considers the future impacts of large-scale integration of battery storage at the residential level. Section 6.6 discusses the ongoing initiatives meant to support the process. Sections 6.7 and 6.8 present some policy and regulatory recommendations, respectively. Section 6.9 summarizes the chapter, highlighting the key role of policy in the integration of residential battery storage. Emanating from these, Section 6.10 proposes future work in this segment.

6.2   STAKEHOLDERS IN THE BATTERY AGGREGATED NETWORK

Battery storage is one of the distributed energy resources (DERs) integrated at the residential level. As mentioned in the previous section, they have been used as a medium to store power supplies from the grid. With the growth of renewable energy resources, particularly rooftop solar PV, there has been a shift in dependency from the grid. The section considers the role of aggregation for voluntary control of battery storage and promoting its integration at the residential level, suggesting policies for the same. Figure 6.1 identifies the major stakeholders who are deriving value from the battery storage integration. The various stakeholders in the aggregated battery storage network include the following:

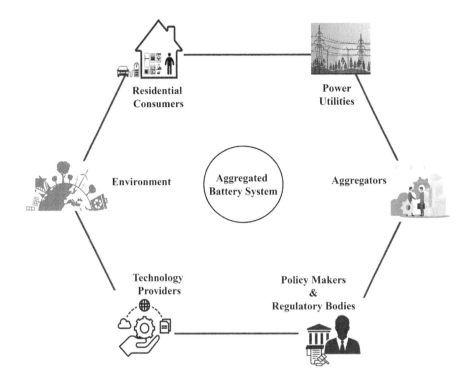

FIGURE 6.1   Stakeholders of an aggregated battery storage system.

- *Residential consumers.* These form a significant element of the battery aggregated network. Beyond the typical advantages, battery storage helps residential consumers save on electricity bills through the increased dependency on rooftop solar PV. In addition, they also benefit from economic incentives for providing DR services, with effective use of battery storage as seen in HEMS or GEBs.
- *Aggregators.* They are responsible for managing the aggregation of battery storage with its concerned beneficiaries, as mentioned below. They aim to derive economic benefits by enabling residential consumers to participate in energy markets through battery storage integrated with rooftop solar PV. By means of advanced algorithms, they can help integrate residential battery storage optimally with the grid in response to real-time grid conditions.
- *Power utilities and grid operators.* Aggregated battery storage helps the power utilities maintain a supply–demand balance with the effective use of DERs. The grid operators include the distribution system operators (DSO) and the transmission system operators (TSO). They are responsible for managing congestion issues on the grid. The residential-level operation involves the active role of DSO to manage the congestion at the distribution level resulting from increased rooftop solar PV integration. By coordinating with the aggregator

and optimally utilizing the battery storage, they can avoid the need for grid reinforcement.

- *Technology providers.* They play a crucial role in developing, supplying, and maintaining the hardware and software components that enable the efficient functioning of aggregated battery systems. The hardware components include battery storage, sensors, charging infrastructure, and other physical components meant for aggregation. The software components, on the other hand, include optimally designed algorithms to control charging/discharging patterns, thermal conditions, and the real-time response behavior of the battery storage. Apart from these, they are responsible for establishing two-way communication interfaces, ensuring interoperability protocols, and addressing privacy issues. Aggregation improves the technological innovation aspect of the services provided by these entities and helps them collaborate with other stakeholders.
- *Environment.* Aggregation can help fulfill decarbonization goals for the environment. The power sector is a major contributor to greenhouse gases, leading to ozone layer depletion and excessive heating of the Earth's atmosphere. The resulting climatic changes are a major concern worldwide. Battery storage contributes to mitigating this challenge by decreasing reliance on fossil fuels through the widespread adoption of rooftop solar PV systems.
- *Policymakers and regulatory bodies.* They are responsible for formulating policies to address the various challenges of battery storage aggregation, as discussed in the following section. Regulators, in this context, may emphasize setting obligations for DR services using battery storage and setting effective tariff designs in order to comply with them, besides incentivizing consumers. The regulators also need to ensure compliance with other aspects of battery storage to manage grid congestion issues due to the increased integration of rooftop solar PV.

6.3   BENEFITS OF BATTERY STORAGE AGGREGATION

Battery storage integrated at the residential level can be controlled through direct scheduling as well as aggregation. For this, the choice of storage is made depending on the consumer's needs and their income level, which are typically categorized as [3]:

- Small battery storage, up to 5 kWh ensures uninterrupted power supplies and provides DR services;
- Medium battery storage, 5 kWh to 15 kWh is used for integrating with rooftop solar PV in order to increase PV self-consumption; and
- Large battery storage, exceeding 25 kWh helps store sufficient power to be less dependent on the grid. In this case, it is ensured that the needs of consumers are effectively satisfied with increased grid independence. Dynamic batteries used in EVs also fall into this category.

Aggregation involves bringing a third-party agent (i.e., an aggregator) as a mediator between the residential consumers and the grid. This requires bidirectional control, facilitating two-way interaction between the battery storage and the grid. On the other hand, direct scheduling is related to the individual-level ownership of battery storage. It involves consumers controlling the charging/discharging patterns by themselves to have sufficient power for their needs. There is little motivation for providing grid services at a voluntary level. Aggregation ensures a voluntary level of action for providing grid services by controlling a large number of batteries. Such a control tries to increase the dependency of residential consumers on solar PV and use it to charge the battery storage for later use. It requires charging to take place during solar hours, with an efficient and smart charging control mechanism. In the absence of this mechanism, charging may happen as soon as the rooftop PV generation exceeds the demand. This results in storage getting fully charged before noon, despite the availability of sufficiently large solar irradiance [6]. Aggregators can implement an optimized charging/discharging scheme to enhance battery storage capacity, fostering cost benefits for consumers [7]. They can also help in managing the excess PV export rate and thus reduce the need for network reinforcements as well as the service costs for the utilities.

For power utilities, the availability of a large storage capacity integrated at the residential level helps facilitate DR, with the availability of sufficient storage capacity. This provides flexibility to consumers for shifting their consumption pattern easily being assured with available capacity. Various provisions are being made for this in intelligently controlled GEBs or HEMS, but consumers, through direct scheduling, show a tendency to be less responsive when delivering DR services. Aggregation can help in getting this done at a voluntary level for the fulfillment of DR obligations to benefit the utilities. It also facilitates ancillary services, frequency containment reserves, voltage regulation, and congestion management at the distribution level [8, 9]. Aggregation can also be used to provide reserve services at the transmission level [9]. Sufficient storage during events of high impact on the grid helps to maintain resilience and reliability [10]. Battery storage aggregation can facilitate a pathway for residential consumers to provide system-level services in addition to fulfilling their individual needs. The use of EV batteries will be even more beneficial as it involves the use of a large number of batteries, ensuring more storage capacity. Though at present they are meant only for mobility, in the coming years they will ensure the availability of more energy, unlike that obtained from a small number of batteries. The government can consider investing collectively in battery storage to provide mobility as well as for household purposes. This voluntary level of control ensures that consumers, utilities, and the environment benefit the most from using the diversified range of batteries made available through the aggregator.

6.4   CHALLENGES OF BATTERY STORAGE AGGREGATION

Battery aggregation is about bringing together various scattered batteries, be they static or EV battery storage. It involves the role of an entity such as an aggregator to link residential consumers and power utilities. This can be done through the use of communication interfaces that facilitate the two-way exchange of information. It turns out to be a significantly directed approach, using various battery storage options at the residential level to serve the system-level needs and help in decarbonization. The technical, economic, and societal challenges in this process are described further in detail.

6.4.1 TECHNICAL CHALLENGES

Considering residential battery storage, the technical complexities of static batteries are few in comparison to those used in EVs. For the latter, the vehicle to grid (V2G) capability needs to be developed in order to use them to discharge power to the grid. These may also be used to share power within the residences as required. However, a challenge arises, as currently only a few EVs have integrated V2G technology. Using this to provide system-level services with the involvement of a large number of actors will be quite complex [8]. There is a need for a developed bidirectional charging mechanism to facilitate power flow from the battery storage, based on either wireless or wired mechanisms. Such an advancement will facilitate effective charging/discharging when connected either to rooftop solar PV or directly to the grid.

Due to a lack of standardized protocols, there is significant inefficiency and induction power transfer loss in the wireless mechanism [10]. The construction of effective charging stations is required to facilitate the use of these residential batteries for providing system-level benefits and, hence, can be used for aggregation. Based on the voltage level, three types of battery chargers need to be developed at present [11]:

- Level 1 or low voltage chargers, which are present in homes and involve slow charging. They require power up to 1 kW to 2 kW;
- Level 2 chargers, which require power in the range of 4 kW to 20 kW; and
- Level 3 or fast chargers with power consumption up to 50 kW, 150 kW, or 350 kW. These charge the vehicle as fast as the speed of a gas station.

Following the lack of robust charging infrastructure, the second shortcoming is the availability of bidirectional communication interfaces. There is a need to ensure protocols for interoperability between the batteries, the charging equipment, and the aggregators to facilitate the use of the system by stakeholders. Protocols such as the open charge point protocol (OCPP) and ISO 61850 form the communication linkage between the battery charging infrastructure and the aggregators. Both static as well as EV batteries need to be coordinated with the aggregator and, thereafter, the aggregator with the system-level utilities [8].

The third issue is the management of rooftop solar PV integrated with battery storage. Uncontrolled charging/discharging patterns of the battery storage, mostly those connected to rooftop solar PV, may lead to network issues at the distribution level [4]. To manage this entire process and, at the same time, ensure the availability of power for system services, is a challenging task. Failing to incorporate these may lead to frequent peak occurrences, harmonic injection affecting the power quality, and a low power factor on the congested network. It also leads to phase and voltage imbalances due to an increase in the charging density over a given period [10]. Ensuring optimal coordination between the residential consumers and the power utilities will be necessary to manage congestion arising from the significant charging/discharging.

The fourth issue is the intricacy of managing the data related to certain aspects, such as the charging duration, the number of battery storage charges during a given period, and unit energy consumption. Additional data are required, especially for the EV batteries, such as the availability hours and route data, as these could power the grid or the residential complexes when parked [10]. The unavailability of battery storage during hours of high solar insolation is an issue. This emphasizes the need for a robust data management system interface for large aggregated networks. Besides, a large number of power converters and control equipment are needed to control the residential battery system for their intended needs effectively. There will be a need to track the availability of battery capacity on a daily basis after the fulfillment of individual-level requirements in order to satisfy the system requirements.

Lastly, frequent charging/discharging poses a concern for the degradation of battery storage. It depends on the amount of charge being drawn, the cycling frequency, and the state of charge (SOC) at which batteries are being operated. This may affect the efficiency of battery storage with a comparatively poor discharge rate. Aggregation of various residential batteries may, in the future, lead to the above-mentioned issues, despite their associated benefits.

6.4.2 ECONOMIC CHALLENGES

The high cost of battery storage reduces the motivation needed for residential-level integration. This may further slow down the smooth and efficient aggregation process. Economies of scale, increased investments, and the penetration of local-level companies can be used as tools to reduce the cost to an optimal level, but a lack of fabrication efficiency in terms of the materials being used in battery storage may overpower the positive effects of these three forces. This is because even though several economic policies may come up to help reduce the upfront cost of the batteries, a lack of fabricating material (i.e., cobalt and nickel reserves) will overpower their impact. In this way, an effort at the market level to help reduce the price may not be as effective.

Secondly, there is a lack of market rules for promoting investments in battery charging infrastructure enabling battery storage to power the grid. There is an ongoing dilemma between automakers and electric utilities regarding over-investment in charging infrastructure for mobile residential batteries [8]. Investment in charging infrastructure, along with using efficient and lightweight semiconductors for efficient battery charging, will lead to increased integration. It can promote the setting up of many battery charging stations, accelerating the process of aggregation. Static batteries are mostly compatible with grid integration, but batteries being used for mobility need to be developed. Presently, it is a challenge for the government to collectively invest in EV battery storage for household as well as mobility needs.

Another crucial challenge is the degradation cost considerations for battery storage. Aggregation will result in batteries getting repeatedly charged and discharged, reducing their life span to less than the conventional duration of about 3–5 years with a high upfront cost [8]. This, to date, has been the reason for less willingness to adopt increased battery storage. Adding to it is the difficulty in deciding the degradation cost of batteries. The debatable issue is deciding upon whom this cost will be levied, the aggregator responsible for the repeated charging/discharging process, or the consumers having ownership of battery storage.

6.4.3 SOCIETAL CHALLENGES

Societal acceptance is a determining factor for the viability of any policy. It ensures the implementation of a technology and harnesses the associated benefits. In this regard, willingness to pay (WTP) is a variable that monetarily quantifies the readiness of residential consumers to invest in battery storage technology and associated aggregated infrastructure [12]. Residential consumers are least bothered about providing system-level services until they are promised identified returns. Apart from this, other factors such as cost considerations, lack of awareness about promising returns on expenditure, the income level of residential consumers, and household characteristics need to be considered. The lack of provisions for coupling different technologies, such as solar plus battery systems or automakers bringing in EV batteries with onboard charging infrastructure, needs to be worked on. This will help to win residential consumers' confidence and boost integration. Instead of bringing different technologies separately with their costs, benefits, or constraints, if they are coupled, consumers will be more interested in investing in aggregate infrastructure.

Another concern is the privacy issue that arises following the need for residential consumers to grant permission to third-party agents to access or control their assets. Possible threats from a cyberattack can lead to severe harm. There is a concern related to aggregators, who, being a third party, may tend to exploit the available storage resources, overlooking the needs of consumers.

6.4.4 REGULATORY CHALLENGES

The lack of appropriate regulatory frameworks can pose a challenge to the effective aggregation of battery storage. A strategically organized aggregated battery storage network ensures the benefit of all the involved stakeholders. The obligation framed should not be overambitious, which otherwise may lead to consumers opposing it. Poor framing of obligations may even provoke interested residential consumers to withdraw from them [13]. There is a need for DR obligations from the aggregated battery storage network to ensure that the involved stakeholders benefit. This involves considering the price-based DR and the incentive-based DR [14]. Price-based DR aims to change electricity consumption based on electricity prices, whereas incentive-based DR encourages consumers to shift their consumption by rewarding them. To get DR services at a voluntary level through battery storage, mostly integrated with solar PV, there is a need to design the tariff structure effectively. This should mostly target the price-based DR, as unlike the incentive-based DR, it is sensitive to residential consumers' willingness to shift their consumption.

There is a lack of load flexibility due to the inability to predict and control consumer consumption patterns. Besides, electricity expenses for residential consumers form a small fraction of their household income. For this reason, they may be less motivated to change their daily routine and use of battery storage, increasing the uncertainty of their contribution to DR. Ineffective tariff design, which is unable to represent the system requirements, the impact residential consumers inflict on the system and absence link between the retail and wholesale prices, proves ineffective to promote DR [15].

6.5   COMPLEMENTARY ASPECTS

This section discusses some of the future implications of battery storage in light of the uptake of rooftop PV worldwide. As understood, battery storage has been used for a long time at the residential level. The increased trend of EV adoption has led to the consideration of using EV battery storage for supplying power to different residences as well as the grid. These, if aggregated through the development of infrastructure and a suitable communication interface, will lead to residential as well as system-level benefits. Despite this, there is a need to consider the impact that increased adoption of battery storage will create and develop policies to manage them effectively.

6.5.1   DISPOSAL ISSUES

The mass integration of battery storage will lead to disposal issues. A few methods and technologies are being developed for recycling. It involves the removal of harmful materials before disposal in landfills and the segregation of essential materials that could be used in the future. Regulators should consider defining certain protocols that are battery-specific for the recycling process due to the differences in the composition of each battery [8]. There is a need to consider efficient management of various attributes of battery storage, such as incorporating an intelligent thermal management system to prevent fire hazards, life span considerations, managing charging/discharging efficiency, and many more, to add to the state of health of the batteries [16].

6.5.2   VICIOUS CYCLE OF PRICE HIKES

Over the past few years, there has been an inclination among residential consumers to integrate solar PV generation. This has further boosted the uptake of battery storage, which in turn may more likely create a vicious cycle of high prices. Residential-level demand forms nearly 30 percent of the total electric power demand [3]. A lot of cost-intensive infrastructure at the distribution level is due to residential customers. Their growing confidence in grid independence with batteries and solar integration will result in a loss of revenue for distribution utilities and a consequent increase in the electricity retail price [16]. This is because only a small cost can be recovered from residential consumers. In the process of this cost recovery, the utilities may levy a lot of costs on a small number of those consumers who are unable to disconnect from the grid. Thus, disconnecting from the grid to save on costs may rather lead to consumers paying high prices. Mass adoption of battery storage and their aggregation may enable the use of grid networks, which can serve as a balance to avoid such situations. There will also be a need for capacity-based tariffs instead of network-based tariffs to combat this issue effectively and prevent complete grid detachment [8].

6.5.3 BATTERY PRICE FLOOR ISSUE

With the subsequent increase in research regarding fabrication efficiency, there are future possibilities of price decline, similar to solar PV modules. There is an ongoing price decline in various parts of the world, but despite this, widely used lithium and cobalt batteries are likely to reach a price floor due to the scarcity of active metals in the future [8]. Moreover, there are only a few reserves of cobalt present in areas with serious geopolitical issues. Due to this, there may not be a significant price decline for battery storage, unlike in the case of solar PV. Reutilization of batteries can pave the way for tackling material scarcity and lowering the battery price floor. EV batteries with sufficient balance capacity can be used as static batteries for household purposes and system-level services.

6.6 GLOBAL EFFORTS AND INITIATIVES

There are several challenges to battery storage as well as future impacts, as mentioned in the prior sections. Suitable policy options and regulatory settings are the need of the hour. Battery storage has been used at the residential level, mostly connected to the grid. Instead, at present, countries around the globe are focusing more on battery storage coupled with rooftop solar PV. This has resulted in cost savings on power consumption and adds value to the environment, reducing the dependency on fossil fuels. These initiatives will justify, in the near future, the significance of introducing policies for the aggregation of residential battery storage. The following section explores:

a. Policies that are currently being implemented at the residential level for battery storage integration;
b. Pilot projects worldwide for the integration of battery storage; and
c. A framework for introducing a market policy for battery storage integration.

6.6.1 ONGOING POLICIES

In the initial phase of rooftop solar PV integration, there were policies such as solar feed-in tariffs (FIT) and net-energy metering (NeM). These controlled the way in which consumers connected their self-generation facilities to the grid and were incentivized for it. As an effect, the integration of rooftop solar PV has increased worldwide, with an annual average growth rate of about 50 percent from 2010 to 2020, and continues to grow fast [4]. As the influx of rooftop solar started impacting the grid, governments worldwide started amending their policies. This section mentions policies for promoting battery storage with rooftop solar PV:

- *PV self-consumption policy.* This policy aims to eliminate the shortcomings of the NeM policy, which was implemented to increase rooftop solar PV penetration. Under the NeM policy, consumers paid only for the difference between the rooftop solar PV generation exported and the power imported from the grid. This helped in lowering their power consumption costs, in some cases even reducing them to zero. The consequent increase in solar exports resulted in congestion on the grid. For instance, in 2015, the Hawaii Public Utilities Commission determined that the NeM model was inadequate for managing the significant growth in rooftop PV and also led to grid issues [17]. PV self-consumption policy involves reducing the solar generation export tariff along with battery storage integration. This helps to manage the volume of generated rooftop solar power being exported and effectively mitigates distribution grid issues [6, 17].
- *Smart export.* This policy sought to provide economic incentives to residential consumers for supplying power to the grid. These incentives include savings in electricity bills along with improvements in the efficiency of energy consumption [18]. This policy involves the use of battery storage as home energy storage systems (HESS), especially with rooftop solar PV generation, to provide DR services. The battery storage should be charged during the day or off-peak hours and discharged during peak hours to export power back to the grid [17]. It is being implemented to help consumers shift to rooftop solar PV in return for benefiting the utilities. Such practices can be incorporated at the residential level with HEMS or GEBs in various countries. According to a recent study on residential consumers in Australia, battery storage systems benefit DR programs and offset consumption during peak hours to provide monetary benefits [19].
- *Enhanced net-energy metering.* This is an advanced form of NeM policy, meant for residential consumers to add non-exporting capacity to the distribution system [17]. In order to increase the value of the conventional NeM policy, this policy emphasizes using battery storage options. The decision is underway to implement time of use (TOU) pricing in the compensation rates, ensuring power supply to the grid by consumers during peak periods.

The policies listed above aim to facilitate the choice and investments of residential consumers in the integration of battery storage. Residential consumers can add to the system at their will in return for the incentives they receive.

6.6.2 PILOT PROJECTS

Various pilot projects are underway worldwide, providing capital support for the adoption of battery storage at the residential level. They are available in the form of capital subsidies and incentives for residential consumers to add value to the system.

TABLE 6.1
Residential battery storage pilot projects in different parts of the world

No.	Region	Initiatives/figures
1	Northern Territory, Australia	Home battery scheme for residential battery storage installation of AU$ 450/kWh
2	South Australia	AU$ 100 million budget for 40,000 residential battery storage
3	New South Wales	AU$ 50 million budget targeted for residential battery storage in Smart Energy Program
4	Australian Capital Territory	AU$25 million budget for 5,000 residential battery storage
5	Germany	220,000 new residential battery storage units are to be introduced in 2022, accounting for a total capacity of 1.2 GW/1.9 GWh
6	California	Installation of 251 MW of residential battery storage in 2022, with an estimate for it to double by 2023

Due to this, there has been an increase in the integration of battery storage. For instance, the integration has increased rapidly, from 0.3 MW installations in 2016 to 3.9 MW installations in 2018 (Smart Electric Power Alliance 2018) [20]. A similar trend is seen in various other countries. In 2020, out of the 3000 MW of total installed battery storage capacity in the USA, 960 MW was for behind-the-meter battery storage capacity. It was divided between the residential and non-residential sectors. Residential energy storage markets are expected to increase by 7.5-fold, bringing 7300 MW of capacity through 2025, many of which will be paired with rooftop solar PV [21].

In Australia, a total of 46,417 battery storage units were integrated at the residential level during the period from 2019 to 2021. In 2021, a total of 34,731 battery storage units were installed, leading to tremendous growth in the installation of battery storage with rooftop solar PV [22]. The development of HEMS or GEBs is meant to enhance the services provided to the grid, in addition to the perks enjoyed by residential consumers. Table 6.1 lists a few pilots in different parts of the world regarding the adoption of residential battery storage in a full-fledged manner [23, 24].

6.6.3 MARKET POLICY FRAMEWORKS

Policy options can be instrumental in promoting the integration of battery storage at the residential level. There were policies to support the uptake of solar PV. Upon being effective gradually, there is a need for policies to promote further the uptake of battery storage, especially when coupled with rooftop PV. But in order to make this process more effective on a wider scale, there is a need to strengthen the market for this storage. Table 6.2 presents a framework to develop market policies [17], which will form a foundational basis to help accelerate the development and integration of residential battery storage.

TABLE 6.2
Framework for market policies regarding residential batteries

Policy category	Parent policy	Definition
Market preparation	Planning	Initializing strategic planning for developing residential battery storage and related permitting standards
	Interconnection	Establishing various requirements for connecting residential energy storage to the grid
	Compensation	Creating mechanisms to compensate residential consumers for using battery storage
	Rate making	Implementing rate mechanisms, such as time of use and demand charges, to enhance the uptake of battery storage
Market creation	Wholesale market	Introducing pilot programs for bringing residential battery storage into the wholesale electricity market through the aggregators
	Mandate	Adopting mandates to incentivize the integration of residential battery storage
Market expansion	Storage funding and incentives	Approving incentives to boost the uptake of residential batteries
	Resilience	Developing an energy resilience policy that includes the use of residential-level storage
	Equity	Addressing the impact of residential storage on disadvantaged communities
	Emissions and life cycle impact	Setting an emission target as well as an end-of-life program, considering the possibility of reutilizing battery storage

6.7   POLICY RECOMMENDATIONS

Policies are the set of ideas guiding the decisions to be undertaken in order to obtain a rational goal. At the residential level, these should promote the aggregation of static as well as EV battery storage. Residential consumers, being at a primitive level, are less motivated to make investments in the power sector. It will be possible only if they get any promising returns. The following policy suggestions can deal with the technical, economic, and social challenges associated with battery storage integration, apart from serving the individual-level requirements. A few policies in this context are as follows:

- *Battery charging control policy.* This policy, from a technical perspective, aims to optimize the charging and discharging of battery storage to ensure that sufficient volumes of electric power are available to all residents. In situations where there is large-scale residential PV penetration for supplying power, it will help mitigate the distribution grid issues by storing the excess PV and further utilizing it. This policy focuses on the control strategy to enable charging the battery storage only when excess power generation exceeds a predefined excess power threshold value. It is defined as a percentage of the solar power rating, considering the difference in the level of consumption of residential consumers [6]. Thus, it may ensure the storage is not fully charged before noon when the generation is at its peak; otherwise, the power will have to be exported to the grid without any need. The capacity to store until the evening hours will also result in maximum rooftop solar power availability for the non-solar hours.

• *Storage tariff.* This policy will incentivize consumers in return for supplying electricity to the grid during peak hours, driven by electricity price algorithms (TOU and real-time pricing). It can be effectively incorporated into HEMS or GEBs to provide DR as well as help the system by providing peak shaving services. Incentives are meant to motivate residential consumers to add value to the system.

• *Price gap widening policy (critical peak pricing).* This tariff policy focuses especially on grid-connected battery storage, considering the difference between peak and off-peak prices to encourage residential consumers to earn revenue. This is earned through energy arbitrage, that is, purchasing electricity during off-peak hours and reselling it when the electricity price is high [25]. The return on investment is significant. This policy requires the allocation of a significant portion of battery storage for storing power imports from the grid, in addition to storing solar power generation.

• *Optimal charging/discharging strategy.* There is a need for the equipment of residential battery storage with intelligent technology. This is to control the frequency of the process to reduce the impact of aggregation on battery degradation and the associated system losses. There will be a need to optimize the charging times, and the flow rate, and simultaneously maintain an optimal level of SOC for residential battery storage.

• *Development of charging infrastructure.* Development is required to bring bidirectional charging infrastructure for battery charging with wired or wireless mechanisms. There is a need to conduct research and testing to maintain the quality standards and their impact on the grid and the battery system.

• *Promotion of wide band gap semiconductor materials.* Materials such as gallium nitride [10] with a wide band gap can be used in battery charging infrastructure to ensure they are energy-efficient, lightweight, and small. This is meant for the charging of residential battery storage and also for use to provide power to the grid or household complexes. Such benefits will result in lower operating and service costs.

• *Bidirectional communication interface and standardized protocols.* It is necessary to establish an efficient bidirectional communication interface, inclusive of smart meters, and implement appropriate protocols for seamless interoperability [8]. These should be among the diverse components of the residential battery aggregated system, which include battery storage, battery charging equipment, aggregators, the power grid, and the concerned utilities. Various technologies, such as Bluetooth, WiFi, and processors like AM62, can be used for effective connectivity [10]. These should be linked with a robust cloud-based data management system.

- *Smart battery charger implementation.* Battery storage chargers need to be equipped with an intelligent algorithm to provide DR services to the grid by alerting consumers to charge their batteries during off-peak hours and discharge them during peak hours, either for household purposes or for mobility. This will help the utilities manage the demand as well as the network usage.

- *Transparency in the aggregation process.* This is to ensure that the entire aggregation process is transparent, inclusive, and cross-sectoral involving the concerned stakeholders. This is to standardize this interconnected process and spread awareness as widely as possible.

- *Incorporation of suitable technology.* Various technologies, such as blockchain and the Internet of Energy (IE) should be used to ensure the privacy and security of data in the battery aggregated system. The communication of crucial data between the aggregator and the network entities can be safeguarded through various gateway systems that are end-to-end encrypted.

6.8   REGULATORY SETTINGS

Setting up rules and regulations for the smooth functioning and viability of any process is essential. These regulations should ensure proper implementation of the policies discussed. At the same time, they should provide safeguards against possible malpractices and monopolistic practices while fulfilling the required obligations. Since residential consumers are involved in this very idea, there will be a need for strong legal protection so they do not become misguided or prone to security issues in any way.

- Regulators can play a significant role in designing the tariff based on the wholesale price and grid contingency management. This includes distribution grid based (localized) considerations that were not included in the wholesale price being designed for large price regions or at the national level. It helps mitigate distribution grid issues and ensures rewards for residential consumers for providing the services [26]. Also, they should consider designing the storage tariff inclusive of geographical conditions related to the variations in peak periods. For instance, hotter countries have peak periods mostly during the afternoon due to the increased use of air conditioners, whereas colder regions have significant peaks during the evening. Consumers should have the freedom to choose the tariff design based on their locations or requirements.

- Regulators can implement an obligation on the power utilities to get DR from residential consumers using battery storage through incentive-based regulation. For this, they should consider guiding the charging intensity of household battery storage by charging the residential consumers based on their contribution to peak load and network usage. This will ensure that consumers improve their load profile. Use of battery storage for DR services, via charging during off-peak hours and discharging during peak hours may lead to a second peak before the off-peak [10]. This needs to be effectively managed.

- There is a need for data protection laws to address the privacy and security concerns of residential consumers. These consumers are not confident about giving control of their devices to any third-party agent, perceiving security risks for their household apparatus [27]. This should be aligned with suitable communication protocols as well, within the aggregated network.
- Regulators should control the extent of storage usage and grid dependency so as to prevent complete detachment from the grid leading to the creation of a 'death spiral' for the consumers relying on the grid. The formulated rules should ensure that the policies do not lead to the distribution utilities being at risk of cost recovery.
- There is a need for regulation to define the suitable size of batteries and develop the reutilization process [28], as well as the recycling units with regard to battery disposal and management. Regulators should also ensure compliance of the aggregators with norms so that they do not take undue advantage of residential consumers by overutilizing the battery storage.
- Regulators should ensure the transparency and inclusiveness of the different stakeholders in the residential battery aggregated process.
- Regulators can make provision for levying the battery degradation cost on the utilities if the charging/discharging rate goes beyond the specified optimal level of SOC.
- There is a need for supervision and regular monitoring of the framework to provide subsidy support for the uptake of battery storage coupled with rooftop solar PV [29].

6.9   CONCLUSION

There has been a gradual implementation of different policy options to promote the integration of rooftop solar, bringing in the need for battery storage at the residential level. This chapter has presented another dimension to the policy recommendation for battery storage integration by highlighting the need for aggregation. Aggregation at this level is meant to develop static and EV battery storage, using them to provide system-level services. In this regard, the benefits of aggregation have been discussed in this chapter, which has tried to enumerate the values that can be added to the major stakeholders including power utilities, residential consumers, and our environment. The hindrances and certain additional aspects as a consequence of battery storage aggregation have also been discussed. Finally, the policy and regulatory recommendations have been presented in this aspect. The focus is not on imposing obligations on residential consumers for the integration, as battery storage is still a costly solution. In various countries, the government is trying to make it a viable option for residential consumers by implementing suitable market frameworks.

Appropriate policy options and regulatory settings will certainly help in the integration of battery storage at the residential level.

6.10 FUTURE WORK

This chapter has considered policy options and regulations for the grid-connected residential system. These include various residences and building complexes, from the rural to the urban level. Future work in this context can consider working on isolated and remote areas not connected to the grid. Deserted places or certain zones in hilly areas may limit people's access to the benefits of various government-implemented policies. In these areas, there is a lack of access to power supplies. The source of power supply, if any, is through rooftop solar PV generation, which, being intermittent, cannot guarantee the regularity of power supply without any storage medium. Therefore, it becomes necessary to work on extending the value of battery storage integration to such residential areas, despite being a less cost-effective option.

REFERENCES

1. W. Matar, "Beyond the end-consumer: How would improvements in residential energy efficiency affect the power sector in Saudi Arabia?" *Energy Efficiency*, vol. 9, pp. 771–790, 2016.
2. Z. Yang, J. Zhang, M. C. W. Kintner-Meyer, X. Lu, D. Choi, J. P. Lemmon, and J. Liu, "Electrochemical energy storage for green grid," *Chemical Reviews*, vol. 111, no. 5, pp. 3577–3613, 2011.
3. S. Agnew and P. Dargusch, "Consumer preferences for household-level battery energy storage," *Renewable and Sustainable Energy Reviews*, vol. 75, pp. 609–617, 2017.
4. A. I. Nousdilis, G. C. Kryonidis, E. O. Kontis, G. A. Barzegkar-Ntovom, I. P. Panapakidis, G. C. Christoforidis, and G. K. Papagiannis, "Impact of policy incentives on the promotion of integrated PV and battery storage systems: A techno-economic assessment," *IET Renewable Power Generation*, vol. 14, no. 7, pp. 1174–1183, 2020.
5. D. Parra, and M. K. Patel, "The nature of combining energy storage applications for residential battery technology," *Applied Energy*, vol. 239, pp. 1343–1355, 2019.
6. S. W. Alnaser, S. Z. Althaher, C. Long, Y. Zhou, J. Wu, and R. Hamdan, "Transition towards solar photovoltaic self-consumption policies with Batteries: From the perspective of distribution networks," *Applied Energy*, vol. 304, pp. 117–859, 2021.
7. B. Diouf and R. Pode, "Potential of lithium-ion batteries in renewable energy," *Renewable Energy*, vol. 76, pp. 375–380, 2015.
8. I. S. F. Gomes, Y. Perez, and E. Suomalainen, "Coupling small batteries and PV generation: A review," *Renewable and Sustainable Energy Reviews*, vol. 126, pp. 109–835, 2020.
9. A. Ramos, M. Tuovinen, and M. Ala-Juusela, "Battery Energy Storage System (BESS) as a service in Finland: Business model and regulatory challenges," *Journal of Energy Storage*, vol. 40, pp. 102–720, 2021.
10. M. S. Mastoi, S. Zhuang, H. M. Munir, M. Haris, M. Hassan, M. Alqarni, and B. Alamri, "A study of charging-dispatch strategies and vehicle-to-grid technologies for electric vehicles in distribution networks," *Energy Reports*, vol. 9, pp. 1777–1806, 2023.

11. S. LaMonaca and L. Ryan, "The state of play in electric vehicle charging services – A review of infrastructure provision, players, and policies," *Renewable and Sustainable Energy Reviews*, vol. 154, pp. 111–733, 2022.

12. S. Sundt and K. Rehdanz, "Consumers' willingness to pay for green electricity: A meta-analysis of the literature," *Energy Economics*, vol. 51, pp. 1–8, 2015.

13. B. Parrish, P. Heptonstall, R. Gross, and B. K. Sovacool, "A systematic review of motivations, enablers and barriers for consumer engagement with residential demand response," *Energy Policy*, vol. 138, pp. 111–221, 2020.

14. A. Asadinejad, A. Rahimpour, K. Tomsovic, H. Qi, and C. F. Chen, "Evaluation of residential customer elasticity for incentive based demand response programs," *Electric Power Systems Research*, vol. 158, pp. 26–36, 2018.

15. J. H. Kim, and A. Shcherbakova, "Common failures of demand response," *Energy*, vol. 36, pp. 873–880, 2011.

16. M. Schwarz, J. Ossenbrink, C. Knoeri, and V. H. Hoffmann, "Addressing integration challenges of high shares of residential solar photovoltaics with battery storage and smart policy designs," *Environmental Research Letters*, vol. 14, no. 7, pp. 074–002, 2019.

17. J. J. Cook, K. Xu, S. Jena, M. S. Qasim, and J. Harmon, "Check the storage stack: Comparing behind-the-meter energy storage state policy stacks in the United States," National Renewable Energy Lab (NREL), Golden, CO (United States), Tech. Rep. NREL/TP-6A20-83045, 2022.

18. B. Zhou, W. Li, K. W. Chan, Y. Cao, Y. Kuang, X. Liu, and X. Wang, "Smart home energy management systems: Concept, configurations, and scheduling strategies," *Renewable and Sustainable Energy Reviews*, vol. 61, pp. 30–40, 2016.

19. N. Al Khafaf, A. A. Rezaei, A. M. Amani, M. Jalili, B. McGrath, L. Meegahapola, and A. Vahidnia, "Impact of battery storage on residential energy consumption: An Australian case study based on smart meter data," *Renewable Energy*, vol. 182, pp. 390–400, 2022.

20. J. B. Twitchell, S. F. Newman, R. S. O'Neil, and M. T. McDonnell, "Planning considerations for energy storage in resilience applications. Outcomes from the NELHA Energy Storage Conference's Policy and Regulatory Workshop," Pacific Northwest National Lab. (PNNL), Richland, WA (United States), Tech. Rep. PNNL-29738, 2020.

21. G. L. Barbose, S. Elmallah, and W. Gorman, *Behind the meter solar + storage: Market data and trends*, Lawrence Berkeley National Laboratory (LBNL), California, 2021.

22. Y. Meng, "Economic analysis for centralized battery energy storage system with reused battery from EV in Australia," *International Conference on Energy, Power and Environmental System Engineering*, vol. 300, pp. 01003, 2021.

23. C. Nichols. (2023, Oct.). A view of the rapidly growing Australian BESS market [Online]. Available: www.idtechex.com/en/research-article

24. J. Spector. (2023, Oct.). Battery firms set to thrive under new California rooftop solar regime [Online]. Available: www.canarymedia.com/articles/batteries

25. B. Zakeri, S. Cross, P. E. Dodds, and G. C. Gissey, "Policy options for enhancing economic profitability of residential solar photovoltaic with battery energy storage," *Applied Energy*, vol. 290, pp. 116–697, 2021.

26. B. E. Lebrouhi, Y. Khattari, B. Lamrani, M. Maaroufi, Y. Zeraouli, and T. Kousksou, "Key challenges for a large-scale development of battery electric vehicles: A comprehensive review," *Journal of Energy Storage*, vol. 44, pp. 103–273, 2021.

27. N. Hossein Motlagh, M. Mohammadrezaei, J. Hunt, and B. Zakeri, "Internet of Things (IoT) and the energy sector," *Energies*, vol. 13, no. 2, pp. 494, 2020.

28. M. Shahjalal, P. K. Roy, T. Shams, A. Fly, J. I. Chowdhury, M. R. Ahmed, and K. Liu, "A review on second-life of Li-ion batteries: Prospects, challenges, and issues," *Energy*, vol. 241, pp. 122–881, 2022.

29. D. P. Brown, "Socioeconomic and demographic disparities in residential battery storage adoption: Evidence from California," *Energy Policy*, vol. 164, pp. 112–877, 2022.

7 Off-Grid Battery-Integrated Residential Systems

Ikram El Haji, Mustapha Kchikach,
Abdennebi El Hasnaoui, and Sanaa Sahbani

7.1 INTRODUCTION

Recent statistics indicate that 733 million people are without access to electricity, while 2.4 billion continue to use harmful fuels daily. The COVID-19 pandemic slowed progress toward universal energy access. Based on the current rate of progress, it is concluded that 670 million people will remain without electricity by 2030, 10 million more than projected last year. Furthermore, about 70 percent of the population in the world lives without energy access, especially in Africa and Asia, and 90 million people who had access to electricity will no longer have the capability to pay for energy basis needs [1]. In addition, fossil fuels are continuously decreasing, while global warming dangers and energy power consumption are increasing due to population and demand growth. As a result, looking for strategies to solve the worldwide energy access issue is extremely important.

Recently, renewable energy systems (RES) have been introduced as a promising solution to overcome the depletion of fossil fuels and the increase in pollution levels through the use of clean and safe sources. Solar photovoltaic (PV), hydro, wind, geothermal, and biomass are some of the most well-known RES [2]. However, the PV system emerges as the most promising technology to supply future energy demands because of its low operating costs, simple installation procedure, and easy maintenance [2]. Yet, to ensure the balance between energy demand and the irregularity of RES production, energy storage is considered as one of the successful solutions [3]. RES are divided into multiple types including off-grid systems and grid-connected systems. On the one hand, the integration of on-grid systems into domestic and industrial life is still limited due to synchronization problems [4]. On the other hand, off-grid renewable energy-based power systems are now known as a suitable solution for rural and islanded electrification locations where grid electricity is not efficient or practical [5]. Thus, the implementation of off-grid systems with energy storage options is a suitable solution to cover population energy needs and help to the energy transition process. An off-grid system is frequently used to power residences, rural areas, and households. It consists of one or two renewable energy as a primary source plus a battery to store energy and balance the household demand and power energy flow during peaks [6–8].

DOI: 10.1201/9781003441236-7

Energy storage systems (ESS) make it possible to benefit from the system production while the load demand is low by storing the energy excess in the battery and converting it to the systems when needed. As previously mentioned, off-grid battery-integrated systems rely on renewable energy as a primary source. Yet, the unpredictability of the power flow of these sources pushes to use other devices that allow highly efficient electrical energy production to be stored in batteries and other ESS.

The intermittence of the system's production, before it is stored, is managed by the integration of power converters and their controls. The objective of power converter controls is to maintain the electricity production at its high peak value during the whole duration of the system's functioning. However, the power converter increases the system's efficiency without handling the excess of energy production during periods of low load demand, and so the use of ESS is crucial for a successful and optimal RES.

This chapter focuses on off-grid systems as RES are recommended for use in developed countries, especially in rural areas where electricity access is hard and expensive for both the government and the general population. Among energy consumption sectors (industrial, transportation, and residential), the residential sector has the highest renewable energy consumption [9]. While connected grid RES continue to face challenges in being formally included in the energy transition process, off-grid systems have emerged as a practical and efficient solution for residence and households [9–13]. ESS are divided into multiple types, namely mechanical, electrochemical, chemical, electrical, and thermal [14]. Several studies have discussed different types and applications of ESS [15, 16]. However, the characteristics of residences and households require an accurate and suitable ESS in terms of price, space, cost, and size. Thus, battery appears as the most suitable ESS to be joined with off-grid systems in households and residences.

In addition, the interest in off-grid battery-integrated systems relies on several factors, including replacement of grid extension, PV, and significant cost reductions due to the growth of off-grid systems utilization [17]. Furthermore, the integration of the battery as an ESS in off-grid systems allows flexible use of the stored energy to power AC and DC loads by the integration of DC-AC and DC-DC converters. From a social prospect, the awareness of consumers regarding the environmental damage when using fossil fuels increases the switching to green energy to guarantee the independence of the power grid [18, 19]. Thus, the development of the off-grid systems by the integration of battery as an ESS, especially in the residential sector, is highly recommended to address this need. The research gap regarding off-grid battery-integrated systems is still open, and challenges and technological advances to develop the efficiency and reliability of these systems are still highly debated [20, 21].

The objective of this chapter is to discuss off-grid battery-integrated residential systems (OGBIRS) with the following aims: (a) to provide an overview of OGBIRS; (b) to discuss frequently used renewable energy systems and reasons for using them; (c) to explain the benefits of using ESs; (d) to set out types of ESS, with a focus on battery types, characteristics, and control; (e) to discuss DC-DC and DC-AC converter roles and controls; (f) to consider the design methodology and software used for simulation; (g) to review real case studies; and (g) to identify some remaining challenges for OGBIRS.

7.2 FREQUENTLY USED RENEWABLE ENERGY SYSTEMS

There are several types of renewable energy systems, such as solar photovoltaic systems, wind energy systems, and small hydroelectric systems. However, the use of some of these systems to power households and buildings might be socially and economically difficult and unfriendly for users. Thus, when considering residential application features and criteria among RES, solar photovoltaic systems are the most used RES for OGBIRS, with wind energy systems in second place [22–25]. Table 7.1 lists factors enhancing the use solar photovoltaic and wind energy systems in OGBIRS.

The architecture of RES is divided into two options according to the load demand and the system need. The first type is a one-stage architecture composed of a power source and one stage of conversion, which in the case of OGBIRS involves inverter, battery bank, and load. The second type of architecture, shown in Figure 7.1, contains two stages of conversion (DC-DC and DC-AC conversion). The two-stage architecture is required more for OGBIRS application than the one-stage architecture. The one-stage architecture consists of converting the DC voltage to an AC voltage without looking for the power source efficiency in most cases. However, the two-stage architecture, by means of the DC-DC conversion features, gives the advantages of controlling

---

TABLE 7.1
Solar and wind energy systems factors

Renewable energy system	Factors
Solar photovoltaic system	Variation in size from small roof-mounted installations with few panels and kilowatts capacity to large-scale installations of several megawatts by utilizing large arrays of solar panels [25]
	The decline in the price of solar panels
	Maturity and the steady improvement of the manufacturing sector
	Most used in off-grid systems applications because of the system's flexibility in developing electric systems [25]
	Suitable for powering households and buildings [26]
	Based on cost comparisons and user-friendliness, solar technology utilization is identified as the best technology now available [24]
Wind energy system	One of the fastest-growing sources of renewable energy in the world
	Cost-effective
	Improvements in the technology involved
	Implemented in developing countries [21]

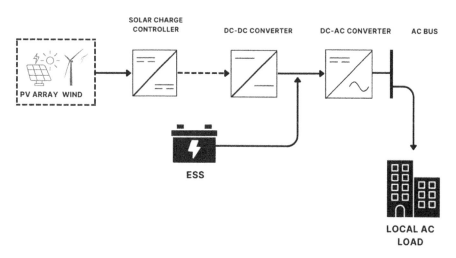

FIGURE 7.1 OGBIRS two-stage architecture.

the voltage level through the integration of boost or buck-boost converters which optimize the OGBIRS design regarding PV panel numbers or wind design. Filter and transformers can be introduced according to the application type and requirement. In most cases, the DC-DC converter is connected to the renewable source through a filter. Its output voltage is converted via the DC-AC converter to meet the AC load requirement. In case of low power demand, the excess of power is stored in batteries directly.

7.3 BENEFITS OF ENERGY STORAGE SYSTEMS

ESS adjust the variability between energy production and customer demand. The use of ESS in RES solves different problems, which include the following [26]:

- *Spinning reserves and short-term backup.* In a typical power system, power plants are coal- and fuel-based. ESs function as spinning reserves to help with power ramp-up as necessary. They can respond in a matter of minutes or seconds, compared to fuel-based systems, which require lengthy startup and ramp-up times.
- *Load balancing and peak shaving.* Using renewable energy systems, electricity is only produced when the used resource is available, which does not always correspond with the time of peak demand. Therefore, this energy mismatch could result in power spillage and losses. This mismatched energy can be moved by the use of ESs in order to meet peak demand and level the load curve [27]. In addition, by teaching consumers about energy efficiency, demand response programs, and energy usage and conservation via ESs, load curve leveling can also be handled from the consumer side.
- *Improvements in power quality.* ESs can also be utilized to improve power quality in areas such as voltage regulation and low voltage ride-through.

- *Smart houses.* Residential ESs are vital for the development of smart homes and energy-efficient structures. The storage of energy can be controlled and managed, which will provide reliable information about the power used and the resulting emissions. These data will serve as ideal benchmarks for reducing energy use and climate control.

7.4   TYPES OF BATTERIES

To ensure reliable energy production for consumers and OGBIRS users, it is crucial to guarantee an excellent feature of ESS: the storage of energy during low demand load allows to feed energy when it is needed during high peak demand. Nevertheless, an inappropriate selection of ESS, responding to OGBIRS user's energy needs, remains inefficient.

ESS include various types Figure 7.2 [28, 29]. In this chapter, only batteries are discussed. Recently, batteries have been acknowledged as the most cost-effective ESS, and they are suitable for operation in a microgrid environment. In batteries, chemical energy is converted into electrical energy, and vice versa, via a series of stacked cells [30]. In addition, the benefits of batteries include fast response times, minimal self-discharge loss, excellent round-trip efficiency, high power and energy densities, mature technology, and ease of commercial availability [26, 30]. These benefits introduce batteries as the most appropriate way to store energy in OGBIRS.

Based on the importance of the integration of batteries in RES in general, and OGBIRS especially, several types of batteries have been discussed in the literature

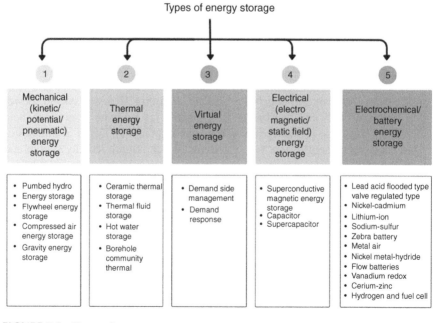

FIGURE 7.2   Types of energy storage systems.

including lead-acid batteries, nickel-based batteries, sodium-sulfur batteries, lithium-based batteries, and VRB flow batteries [31, 32]. Among these types, the lead-acid battery is the most developed, and it is widely used in both industry and society. However, according to some studies [31, 33], Li-ion, Pb-acid, and VRB are the most effective technologies for RES integration and grid energy storage.

- *Lead-acid (L/A) batteries* appeared in 1860, defined as the most popular rechargeable large format batteries on the market at present. Significant technological advancements have been accomplished in this type of battery such as the creation of sealed L/A batteries that are "spill-proof" and may be used in any physical configuration, including upside-down, sideways, and upright [32]. Moreover, lead-acid batteries have impacted power system deployment throughout the past few years due to their durability and cost-effectiveness. The lack of depth of discharge is their main drawback [34, 35].
- *Li-ion batteries* have been widely used in utility grid applications due to their minimal maintenance requirements, high level of safety, and characteristics of large volume and energy density. Their main drawback is that they are not economically viable. Li-ion batteries are very sensitive to cell stability and capable of causing major issues in the event of an impairment. However, Li-ion solutions include complex control mechanisms that monitor cell performance and adjust the operating temperature for the energy storage device in order to function optimally [26, 34].
- *Vanadium redox (VRB) flow batteries*, along with *zinc-bromine (ZnBr) batteries*, are the two main types of flow batteries in the early stages of commercialization and demonstration. The electrolytes can be used indefinitely because each charge/discharge cycle returns the solutions to their initial states, resulting in a very long lifetime for VRB of over 10,000 cycles or above ten years [32]. Since electrolytes are stored externally, the key benefits of VRB batteries are that the power and energy components are easily scalable and may be determined independently. The disadvantage of VRB, as compared to other batteries, is the low specific energy and energy density. As a result, zinc-bromine (ZnBr) has been introduced as an alternative to VRB. ZnBr has a greater specific energy than VRB, but it is less developed than VRB technology in terms of batteries. Polysulfide bromide is another major type of flow battery [26, 32].
- *Sodium-sulfur (NaS) batteries* are typically made of liquid sodium (Na) and sulfur (S) and operate at a high temperature of about 300 °C. These types of batteries have reasonable prices, and they are constructed with low-cost components. In addition, sodium is a low-cost alternative to lithium that is available in different regions all over the world. NaS batteries are widely used in ESs due to their high energy density, charge/discharge efficiency, and long cycle life. This cell chemistry is anticipated to be feasible for energy storage applications such as load balancing, emergency power supplies, and uninterruptible power supplies in terms of scale and cost [36].
- *Nickel-based batteries* include two subtypes: nickel cadmium (NiCd) and nickel metal hydride (NiMH) [29]. The advantages of nickel-based batteries include fast discharging cycles, reduced cost per cycle, long life, and suitability

for renewable applications. Their main drawbacks are cost, the toxicity of cadmium, the memory effect, high exothermicity, and need for continued maintenance [26].

7.5 CHARACTERISTICS OF BATTERIES

Storage devices, such as batteries, have an essential role in renewable ESS. However, unless an appropriate storage device is selected, the role and benefits of the ESS remain limited. Thus, to ensure a good selection of ESS, the storage device is characterized by five features [37]:

- *Cycle efficiency* is the ratio of the amount of energy output during discharging to the amount of energy input during charging. For ESS, a high cycle efficiency is a desirable characteristic. A cycle efficiency close to 100 percent indicates that less energy is lost during charging and discharging cycling [38].
- *Cycle life* represents the maximum number of charging and discharging cycles that an ESS element can perform before its capacity drops below a specific percentage. When the cycle life is over, the ESS elements need to be changed. When selecting the battery type, long cycle life is recommended [37].
- *Self-discharge rate*, or leakage, is a measure of the speed at that an ESS element loses its energy even if there is no current consumed by a load. For long-term energy storage, low self-discharge ESS elements are preferred.
- *Energy and power density* are defined as the maximum energy storage per volume or weight, and the maximum power rating per volume or weight, respectively. High energy and power densities are important for applications where volume and weight are critical constraints [38].
- *Capital cost* is important, as ESS are introduced to help RES to meet the load and power requirements. To ensure the optimization of OGBIRS, a low-cost storage device is preferable.

7.6 BATTERY CHARGE CONTROLLERS

The purpose of using a battery charge controller is to enhance the battery's state of charge and to control effectively its overcharging and discharging state, in addition to increasing the battery's life and charging time. Several battery controllers have been discussed in the literature, especially for lead-acid batteries and Li-on batteries [39, 40]. Battery charge controllers are divided into classic or passive controllers and optimal charge controllers that do not require a mathematical systems model.

There are many types of classic battery charge controllers, namely constant voltage (CV), constant current (CC), two-step charging (CC-CV) which combines (CV) and (CC), pulse charging (PC), negative pulse charging (NPC), trickle charge or taper-current (TC), and float charge (FC). The optimal charge controllers newly developed for battery charge controlling are fuzzy logic control (FLC) and model predictive control (MPC) [41]. Classic and optimal charge controller types can be described as follows.

- *CV.* This method consists of applying a constant voltage on the battery's terminals. At the first stage of charging the current charge is high, while it decreases when the charge voltage reaches its limit [40].
- *CC.* Using this technique, the battery is charged with a constant current. To predict the over-current of the initial charge, the current is limited [42]. The benefit of this technique is that the charging time and SOC can be easily calculated based on the voltage value, which depends on the charging current. However, due to the uncontrolled voltage, the following drawbacks are listed when using a CC battery charge controller: battery overcharging, a rise in temperature, and a reduction in battery life [40].
- *CC-CV.* A CC-CV battery charge controller is a combination of the two previous cited battery charge controllers, CC and CV. This technique is composed of two steps: the first step consists of applying the CC charge controller until the battery reaches a certain preset charging voltage, and in the second step the charging voltage is held constant until the current is reduced to a preset minimum current value [39, 40].
- *PC.* This type of battery charge controller consists of applying a pulse current to the battery periodically, which allows the recharge and discharge of the battery periodically [40]. The pulse charging controller reduces the polarization in order to predict the rising temperature. Nevertheless, pulse charging battery control is complex compared to CC and CV [42].
- *NPC.* NPC battery charge controllers are an improvement on PC battery charge controllers. They include the same pulse technique, but the improvement consists of applying a negative pulse to provide a discharging pulse and a reset time periodically [43].
- *TC.* This technique relies on the battery charging requirement without controlling the current or the voltage. Its main advantage is its low cost; however, it produces noise (harmonics and ripples) that might be harmful when applied to batteries unless a filter is used [44].
- *FC.* The aim of this technique is to ensure that the battery remains fully charged indefinitely by managing the self-discharge of the battery. In addition, this technique allows the increase of the battery's lifetime [41].
- *FLC.* In this technique the battery is charged in two steps. In the first step, a high current is employed to reach 70 percent of the battery's total capacity. In the second step, the current is decreased exponentially while the voltage is held constant [40].
- *MPC.* MPCs are the most popular battery charge controllers for several reasons. The first advantage of this method is its performance in constraint and non-linearity handling in addition to its applicability to a wide range of industrial cases. Its robustness and stability regarding noise also contribute to its widespread use [39, 40].

The battery charge controller is a critical and crucial issue in OGBIRS. Good battery storage management will enhance the efficiency of the whole system. For this reason, it is necessary to develop battery management control carefully in light of the drawbacks and limitations listed above.

7.7 CONVERTERS

Converters are a key element in different sectors that include RES. A great deal of research has discussed different topologies of converters. However, in this chapter, our focus is on DC-DC and DC-AC converters as the most frequently used power converters in OGBIRS. The use of converters in RES generally, and OGBIRS especially, is crucial. They have the ability to convert the voltage or current in terms of forms and levels, which is highly beneficial for intermittent power sources like renewable energy sources, which provide low and variant output power that cannot meet always load requirements and user needs. The use of converters in such cases helps to convert the output power so that it matches, in terms of form and level, the load requirements. Thus, for OGBIRS, DC-DC boost converters will enhance the system's efficiency, and DC-AC converters will convert the voltage's form to meet the AC load requirements. In the following section, DC-DC and DC-AC converters are further explained.

7.7.1 DC-DC CONVERTERS

DC-DC converters are divided into types, isolated and non-isolated [45]. Recently, the most used type is non-isolated converters, thanks to their efficiency and simplicity [46]. Isolated converters are also called transformer-based converters because of the existence of a transformer in this type of circuit. The main advantage given by transformer-based converters is the high gain. Nevertheless, the use of a snubber to reduce voltage spikes in the switches of this type of converter remains crucial, which increases the complexity of the converter and the system as a whole [47].

Boost, buck, and buck-boost converters constitute the typical types of non-isolated DC-DC converters. Boost converters, as their name indicates, provide an output voltage with a higher level than their input voltage, as shown in Figure 7.3. However, the boost converter emerges as the most commonly used converter in RES. The feature of boosting the input voltage level to the highest level, by means of a duty cycle, enhances the RES's efficiency. Unlike the boost converter, the buck converter was introduced as a DC-DC chopper, and its role consists of converting the voltage level to the lowest level while the form remains the same. The buck-boost converter

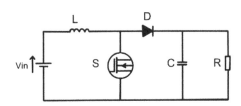

FIGURE 7.3   Boost converter circuit.

combines features of boost and buck converters. Despite their advantages, classical converters still have some limitations, such as high voltage and current ripple, that affect the energy quality in RES and OGBIRS. Therefore, a number of studies have been conducted to overcome these drawbacks. Multilevel converters have recently emerged as a solution to the problems associated with the classic types of converters; they are constituted by a combination of one or two converter types on parallel or series. Reducing current and voltage ripples is one of the advantages provided by multilevel converters that improve the energy quality in RES and OGBIRS [47–51].

7.7.2 DC-DC CONVERTER CONTROL

The output power of renewable energy sources is maximized by the use of controllers of DC-DC converters. In recent years, several controllers have been introduced to produce the highest efficiency for boost converters and RES, by searching for the maximum power point (MPPT) of the renewable energy source.

MPPT techniques can be categorized into classic and developed methods. Recently, AI methods have emerged to improve the classic MPPT techniques. Of the various MPPT techniques discussed in the literature, this chapter considers the perturb and observe (P&O), incremental conductance (IC), fuzzy logic (FL), particle swarm optimization, and moth-flame optimization algorithms [52]. A new MPPT technique based on an arithmetic optimization algorithm for (OGBIRS) is discussed in [53, 54].

- *Perturb & observe (P&O)*. This MPPT technique is based on the determination of the photovoltaic system's output power through the observation of the current voltage and current values, which are then compared to the previous point current and voltage. During this comparison, a series of perturbations are introduced in the voltage or current of the PV panel until the detection of its maximum power point.
- *Incremental conductance (IC)*. This method has been developed to overcome P&O's limitations, which are mainly fluctuations under irradiation and temperature changes, thereby improving the PV system's efficiency by reducing power loss. The basic principle of this method is to compare the incremental conductance and the instantaneous value of the PV to reach the MPPT.
- *Fuzzy logic (FL)*. This type of controller has a number of advantages over traditional approaches: precise mathematical models are not required; it accepts imprecise inputs; it handles non-linearity; and it is more reliable than traditional non-linear controllers. The principle of this method consists of four steps: fuzzification, rules, interference engine, and defuzzification [48].
- *Particle swarm optimization*. The PSO algorithm's basic principle was influenced by observations of animal social behavior, such as flocking birds and schooling fish. The PSO's quick convergence and simple implementation make it suitable for use in MPPT methods [52, 55].
- *Moth-flame optimization algorithm*. MFO, a meta-heuristic optimizer, was developed from the idea that a moth's transverse orientation allows for automated night navigation. Moths take flight at a fixed angle in the direction of the moonlight. If the "flames" of the light source are far away, the moths will

fly in a straight line. They pursue a particular course to make use of the global minimum solutions in the event of closed non-uniform flames or lights [56].

*   *Arithmetic optimization algorithm.* AO draws its main inspiration from how the arithmetic operators (division, multiplication, addition, and subtraction) behave in problem solving. It uses a four-level hierarchy to address optimization issues and locate the global optimal solution without having to compute the derivatives. Results in [53] show the performance of the AO method in terms of charging battery efficiency.

7.7.3 DC-AC CONVERTERS

As with DC-DC converters, the classification of DC-AC converters depends on several factors such as the number of power processing stages (single-stage or multistage), transformer or transformer-less topologies, levels of design, and switching methods [57]. The main role of DC-AC converters, called also inverters, is to convert DC voltage to AC voltage. In OGBIRS, the inverter is used to convert the DC voltage form coming from the battery to an AC form in order to feed AC loads. The presence of an inverter in OGBIRS is vital because most consumers use loads that require AC voltage or current source. However, there is ongoing debate on how to improve inverters; several types have been introduced in the literature in order to provide accurate power quality for REs [58, 59]. Furthermore, some studies aim to add the feature of voltage boosting to the inverter, which is extremely beneficial for OGBIRS [60]. The combination of the inverter's features and boost converter advantages serves to optimize the design approaches and system space for OGBIRS.

7.8 DESIGN METHODOLOGY

In recent years, several research papers have studied the integration of renewable energy systems in rural and islanded areas in addition to households so as to guarantee independence in terms of energy production and consumption. The integration of off-grid systems in such areas not only provides advantages within the energetic and electrical fields but also affects the economic and demographical sides of these areas [61]. However, to study the feasibility of integrating renewable energy systems such as OGBIRS, a number of steps should be considered, including load requirement, geographical site, and component sizing (Figure 7.4). System sizing is particularly critical; as it is the key element that ensures good or bad functioning of the OGBIRS. The design process must therefore be precise and accurate.

Component sizing can follow either of two methodologies. The first (traditional) method sizes the system based on basic equations, while the second methodology uses software tools for that purpose [62].

7.8.1 LOAD REQUIREMENT

Based on the customer's needs, load requirements are defined. A list of DC and AC loads, with their operational hours in the household or in the area where off-grid systems are, is crucial to identify the electrical power needed to feed the building.

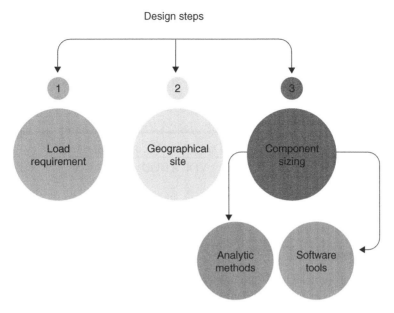

FIGURE 7.4 Design steps.

The load requirement over a year is predicted through the determination of daily energy consumption. The wattage specification is obtained from the specification of the load's features [63].

7.8.2 GEOGRAPHICAL SITE

In order to have an efficient off-grid system it is vital to consider the geographical site while designing the systems. The feasibility of the integration of off-grid systems in energy transitions depends to a great extent on the location of the chosen area or house. This key factor helps to identify the available energy sources based on the site's meteorological and geographical specifications so that the most suitable energy source for the off-grid system power feeding can be selected. In addition, by referring to the NASA Surface Meteorology and Solar Energy website, solar irradiation and wind speed are extracted [5, 12, 64]. Based on the annual extracted data, average daily data are calculated to design the system for daily power feeding.

7.9 COMPONENT SIZING

7.9.1 ANALYTIC METHODS

The theoretical sizing of the off-grid system is detailed in [63] and summarized in this subsection. To guarantee the system's safety, most of the components are usually oversized by 30 percent.

7.9.2 Solar Charge Controller Rating

Solar charge controller rating =

$$\frac{\text{Total short circuit current of PV array}\left(\text{number of parallel string} \times \text{module Isc}\right) \times 1.3}{\text{Temperature derating factor}}$$

7.9.3 Battery Capacity

$$\text{Battery capacity}\left(\text{Ah}\right) = \frac{\text{Daily Wh} \times \text{No. of days of storage}}{\text{VDC} \times \text{DOD} \times \text{CAF} \times \text{CDF}}$$

where VDC is the DC system voltage, CAF is the capacity appreciation factor, DOD is the depth of discharge, and CDF is the capacity degradation factor.

7.9.4 Inverter Sizing

Inverter sizing is crucial in off-grid systems and is the key component of energy production in any off-grid system. When sizing the inverter, a number of elements should be considered, including the AC load, the DC operating voltage of the system, the output voltage, the variation in output waveform, and the frequency. The inverter's capacity should be oversized by 25 to 30 percent for safety.

$$\text{Inverter capacity}\left(\text{VA}\right) = \frac{\text{Total load in Watts}}{\text{Temperature derating factor} \times \text{power factor}}$$

$$\text{Maximum input current to the inverter} = \frac{\text{AC load}\left(\text{W}\right)}{\text{DC voltage}\left(\text{lowest inverter cut} - \text{off voltage}\right)}$$

7.10 SOFTWARE TOOLS

Recently, a number of simulation software tools have been developed to optimize renewable energy systems. In some research papers, software tools are categorized according to whether they are open source or not [25]. However, in other papers, software tools for renewable energy systems are divided into four categories: open architecture research tools that allow changing algorithms and component system interaction (TRNSYS); prefeasibility tools for component sizing and financial analysis of the system (RETScreen); sizing tools that provide the optimal size and the energy flow of the system's component (HOMER); and simulation tools where the user provides specifications of each component so as to have the behavior of the system (HYBRID2) [65].

7.10.1 Transient System Simulation (TRNSYS)

Launched in 1975 by the University of Wisconsin and the University of Colorado in the USA, TRNSYS software was initially developed for thermal simulation systems.

However, TRNSYS has since been upgraded to include various systems involving solar and hybrid systems. The efficiency of this software consists of giving the user the ability to simulate and evaluate the systems with great precision during the time [25, 65, 66].

7.10.2 RETScreen

RETScreen is a clean energy management software tool that enables low-carbon planning, implementation, monitoring, and reporting of systems. It was developed by the government of Canada [67] and is generally used to study the feasibility of renewable energy systems including PV applications, off-grid and on-grid. The main drawbacks of this software tools are that it does not consider the effect of temperature on PV systems, that it does not allow the import of time series data, and that data sharing can be problematic [65].

7.10.3 HOMER

The Hybrid Optimization Model for Electric Renewables (HOMER) was created in the USA by the National Renewable Energy Laboratory (NREL) for the purpose of designing and optimizing microgrid and distributed energy sources [65]. This software is used by more than 250,000 users and developed in over 190 countries. It provides robust solutions to build optimal and maximal cost-effective hybrid power systems. In addition, it has several advantages [68]: it combines engineering and economics in the same module; it determines the most cost-effective solution; and it provides real-world performance simulation and optimized design.

7.10.4 HYBRID2

HYBRID2 was launched in 1996 by the Renewable Energy Research Laboratory of the University of Massachusetts Amherst, USA as an upgrade of HYBRID1, which was developed in 1994. An easily manageable software package for users, HYBRID2 facilitates the evaluation and design of off-grid systems [25, 65, 69].

From this overview of the software tools and the information in Table 7.2 [65], it is concluded that HOMER is the most suitable software tool for studying the feasibility of techno-economic battery off-grid systems in residences, households, and rural areas.

7.11 CASE STUDIES

West China

A case study of a village in West China was conducted by [64] to analyze the feasibility of implementing a hybrid renewable energy system with a battery. The system was studied by following the aforementioned design steps. The simulation was conducted using HOMER software. The optimization results showed that the proposed system generates 322,158 kWh/year, which gives independence to the whole village. The

TABLE 7.2
Software tools used in OGBIRS design: Required inputs and outputs

Software tool	Required inputs	Outputs
TRNSYS	Meteorological data Models from own library	Simulation of dynamic behavior of thermal and electrical energy system
RETScreen	Climate project Product Hydrology databases	Technical, financial, and environmental analysis Sensitivity and risk analysis Energy efficiency Cogeneration
HOMER	Load demand Component details with cost Constraints system control Emission data	Optimal sizing Net present cost Cost of energy Capital cost Capacity shortage Excess energy generation Renewable energy fraction Fuel consumption
HYBRID2	Load demand Power system component details Financial data	Technical analysis Sizing optimization Financial evaluation

system consists of 104 kW PV modules, three wind turbines of 10 kW, a 50 kW BDG, a 99 kW converter, and 331 kWh storage batteries.

Tazouta, Morocco

A case study was conducted [70] to propose the design, simulation, and optimization of a stand- alone photovoltaic system in Tazouta, which is a village located in the Atlas Mountains in Morocco. The main purpose of the study was to provide electrical energy based on a renewable source for a rural house. The design was based on the above-mentioned design steps; the software packages used to verify the feasibility of the study were PVsyst and HOMER PRO. The results show that the system composed of 1080 Wp of PV module capacity and 670 Ah of battery storage has the ability to meet kWh/d of daily electric load.

Dina Farms, Egypt

In the Dina Farms optimization system, sizing and design were based on two steps. First of all, the design procedure was based on the watt-hours consumption calculation. Secondly, PVsyst was used for simulation. The results give two options for solar panels: 260 W and 450 W PV. Both PV panels can meet the load requirements. Nevertheless, the 260W PV panel is more beneficial than a 450W PV panel when considering unused energy, lower power loss in standard conditions, and a higher performance ratio, whereas the 450W PV panel is more attractive than the 260W PV when the goal of sizing is high PV conversion efficiency, high value of energy produced per year, and low area [71].

Engineering College Bikaner, India

In order to feed load requirements into the mechanical department office at Engineering College Bikaner, an OGBIRS was designed using PVsyst software. The study proved that solar panels can provide 1143.6 kWh; due to system power loss, the user benefits from 1068.12 kWh. However, the load requirement goes up to 1086.24 kWh, which is close to the system's supplied energy [72].

7.12 CHALLENGES

In off-grid systems, batteries are most commonly used as ESs for residences and households. This technology is vital to ensure energy feeding to rural and islanded areas. However, despite the steady reduction of renewable energy system device prices, cost is still a barrier to implementing OGBIRS over a wide range. The main challenge in renewable energy systems with the ESS option is ensuring high efficiency and power quality at a reasonable and accessible price. Public awareness and investment in developing energy storage systems are crucial to enhance development in this field. Furthermore, regardless of the high cost of the renewable energy system with energy storage option, the system's size is still an important factor that does not encourage users to look for clean and storable energy, especially in cities where space is limited. Thus, optimizing the system's real size by introducing micro components with good efficiency is a promising solution to this problem. An integrated PV unit-battery has been discussed in the same context [73].

7.13 CONCLUSION

This chapter has presented an overview of off-grid battery-integrated residential systems, providing several reasons to consider these systems as an efficient way to involve the residential sector in the energy transition process. The types of OGBIRS architecture have been introduced and detailed, along with the roles, types, characteristics, and controllers of batteries. The chapter has also focused on converters and MPPT techniques as essential components in developing an efficient OGBIRS. A detailed design methodology for optimizing the OGBIRS has been explained, as have some emerging software tools used in OGBIRS optimization, their features, their required input, and their advantages. An international example of OGBIRS has been provided in rural areas with the same design steps.

To conclude, OGBIRS continues to face many challenges. Arriving at an efficient integration of OGBIRS into the lives of consumers will require further development of renewable energy systems in general, and OGBIRS in particular.

REFERENCES

[1]   World Bank Group, "Tracking SDG 7 – The Energy Progress Report 2022," *World Bank*. www.worldbank.org/en/topic/energy/publication/tracking-sdg-7-the-energy-progress-report-2022 (accessed May 18, 2023).

[2] A. Bharatee, P. K. Ray, B. Subudhi, and A. Ghosh, "Power management strategies in a hybrid energy storage system integrated AC/DC microgrid: A review," *Energies*, vol. 15, no. 19, p. 7176, Jan. 2022, doi: 10.3390/en15197176

[3] A. Berrada, A. Emrani, and A. Ameur, "Life-cycle assessment of gravity energy storage systems for large- scale application," *J. Energy Storage*, vol. 40, p. 102825, Aug. 2021, doi: 10.1016/j.est.2021.102825

[4] K. Seifi and M. Moallem, "Synchronization and control of a single-phase grid-tied inverter under harmonic distortion," *Electronics*, vol. 12, no. 4, Art. no. 4, Jan. 2023, doi: 10.3390/electronics12040860

[5] M. L. Kolhe, K. M. I. U. Ranaweera, and A. G. B. S. Gunawardana, "Techno-economic sizing of off-grid hybrid renewable energy system for rural electrification in Sri Lanka," *Sustain. Energy Technol. Assess.*, vol. 11, pp. 53–64, Sep. 2015, doi: 10.1016/j.seta.2015.03.008

[6] E. I. Come Zebra, H. J. van der Windt, G. Nhumaio, and A. P. C. Faaij, "A review of hybrid renewable energy systems in mini-grids for off-grid electrification in developing countries," *Renew. Sustain. Energy Rev.*, vol. 144, p. 111036, Jul. 2021, doi: 10.1016/j.rser.2021.111036

[7] Y. Zhang, T. Ma, and H. Yang, "A review on capacity sizing and operation strategy of grid-connected photovoltaic battery systems," *Energy Built Environ.*, vol. 5, pp. 500–516, Apr. 2023, doi: 10.1016/j.enbenv.2023.04.001

[8] R. Madurai Elavarasan *et al.*, "The untold subtlety of energy consumption and its influence on policy drive towards Sustainable Development Goal 7," *Appl. Energy*, vol. 334, p. 120698, Mar. 2023, doi: 10.1016/j.apenergy.2023.120698

[9] A. Zahedi, H. A. Z. AL-Bonsrulah, and M. Tafavogh, "Conceptual design and simulation of a stand-alone Wind/PEM fuel cell/hydrogen storage energy system for off-grid regions, a case study in Kuhin, Iran," *Sustain. Energy Technol. Assess.*, vol. 57, p. 103142, Jun. 2023, doi: 10.1016/j.seta.2023.103142

[10] C. Mokhtara, B. Negrou, N. Settou, B. Settou, and M. M. Samy, "Design optimization of off-grid hybrid renewable energy systems considering the effects of building energy performance and climate change: Case study of Algeria," *Energy*, vol. 219, p. 119605, Mar. 2021, doi: 10.1016/j.energy.2020.119605

[11] B. K. Das, M. A. Alotaibi, P. Das, M. S. Islam, S. K. Das, and M. A. Hossain, "Feasibility and techno-economic analysis of stand-alone and grid-connected PV/Wind/Diesel/Batt hybrid energy system: A case study," *Energy Strategy Rev.*, vol. 37, p. 100673, Sep. 2021, doi: 10.1016/j.esr.2021.100673

[12] O. D. T. Odou, R. Bhandari, and R. Adamou, "Hybrid off-grid renewable power system for sustainable rural electrification in Benin," *Renew. Energy*, vol. 145, pp. 1266–1279, Jan. 2020, doi: 10.1016/j.renene.2019.06.032

[13] J. Li, P. Liu, and Z. Li, "Optimal design and techno-economic analysis of a solar-wind-biomass off-grid hybrid power system for remote rural electrification: A case study of west China," *Energy*, vol. 208, p. 118387, Oct. 2020, doi: 10.1016/j.energy.2020.118387

[14] A. Emrani, A. Berrada, and M. Bakhouya, "Optimal sizing and deployment of gravity energy storage system in hybrid PV-Wind power plant," *Renew. Energy*, vol. 183, pp. 12–27, Jan. 2022, doi: 10.1016/j.renene.2021.10.072

[15] E. Hossain, H. M. R. Faruque, M. S. H. Sunny, N. Mohammad, and N. Nawar, "A comprehensive review on energy storage systems: Types, comparison, current scenario, applications, barriers, and potential solutions, policies, and future prospects," *Energies*, vol. 13, no. 14, p. 3651, Jan. 2020, doi: 10.3390/en13143651

[16] S. Koohi-Fayegh and M. A. Rosen, "A review of energy storage types, applications and recent developments," *J. Energy Storage*, vol. 27, p. 101047, Feb. 2020, doi: 10.1016/j.est.2019.101047

[17] P. Ortega-Arriaga, O. Babacan, J. Nelson, and A. Gambhir, "Grid versus off-grid electricity access options: A review on the economic and environmental impacts," *Renew. Sustain. Energy Rev.*, vol. 143, p. 110864, Jun. 2021, doi: 10.1016/j.rser.2021. 110864

[18] S. Misak and L. Prokop, "Off-grid power systems," in *2010 9th International Conference on Environment and Electrical Engineering*, vol. 2010, pp. 14–17, 2010, doi: 10.1109/EEEIC.2010.5490003

[19] S. Mandelli, J. Barbieri, R. Mereu, and E. Colombo, "Off-grid systems for rural electrification in developing countries: Definitions, classification and a comprehensive literature review," *Renew. Sustain. Energy Rev.*, vol. 58, pp. 1621–1646, May 2016, doi: 10.1016/j.rser.2015.12.338

[20] I. E. Atawi, A. Q. Al-Shetwi, A. M. Magableh, and O. H. Albalawi, "Recent advances in hybrid energy storage system integrated renewable power generation: Configuration, control, applications, and future directions," *Batteries*, vol. 9, no. 1, p. 29, Jan. 2023, doi: 10.3390/batteries9010029

[21] B. Li, Z. Liu, Y. Wu, P. Wang, R. Liu, and L. Zhang, "Review on photovoltaic with battery energy storage system for power supply to buildings: Challenges and opportunities," *J. Energy Storage*, vol. 61, p. 106763, May 2023, doi: 10.1016/ j.est.2023.106763

[22] S. M. Aarakit, J. M. Ntayi, F. Wasswa, M. S. Adaramola, and V. F. Ssennono, "Adoption of solar photovoltaic systems in households: Evidence from Uganda," *J. Clean. Prod.*, vol. 329, p. 129619, Dec. 2021, doi: 10.1016/j.jclepro.2021.129619

[23] T. M. Qureshi, K. Ullah, and M. J. Arentsen, "Factors responsible for solar PV adoption at household level: A case of Lahore, Pakistan," *Renew. Sustain. Energy Rev.*, vol. 78, pp. 754–763, Oct. 2017, doi: 10.1016/j.rser.2017.04.020

[24] E. Nyholm, J. Goop, M. Odenberger, and F. Johnsson, "Solar photovoltaic-battery systems in Swedish households – Self-consumption and self-sufficiency," *Appl. Energy*, vol. 183, pp. 148–159, Dec. 2016, doi: 10.1016/j.apenergy.2016.08.172

[25] M. F. Akorede, "2 – Design and performance analysis of off-grid hybrid renewable energy systems," in *Hybrid Technologies for Power Generation*, M. Lo Faro, O. Barbera, and G. Giacoppo, Eds., in Hybrid Energy Systems. Academic Press, 2022, pp. 35–68, doi: 10.1016/B978-0-12-823793-9.00001-2

[26] A. H. Fathima and K. Palanisamy, "8 – Renewable systems and energy storages for hybrid systems," in *Hybrid-Renewable Energy Systems in Microgrids*, A. H. Fathima, N. Prabaharan, K. Palanisamy, A. Kalam, S. Mekhilef, and Jackson. J. Justo, Eds., in Woodhead Publishing Series in Energy. Woodhead Publishing, 2018, pp. 147–164, doi: 10.1016/B978-0-08-102493-5.00008-X

[27] P. Prajof, S. Mohan Krishna, J. L. Febin Daya, U. Subramaniam, and P. V. Brijesh, Eds., *Smart Grids and Microgrids: Technology Evolution*. Wiley, 2022.

[28] A. Emrani, A. Berrada, A. Arechkik, and M. Bakhouya, "Improved techno-economic optimization of an off- grid hybrid solar/wind/gravity energy storage system based on performance indicators," *J. Energy Storage*, vol. 49, p. 104163, May 2022, doi: 10.1016/j.est.2022.104163

[29] F. A. Bhuiyan and A. Yazdani, "Energy storage technologies for grid-connected and off-grid power system applications," in *2012 IEEE Electrical Power and Energy Conference*, pp. 303–310, Oct. 2012, doi: 10.1109/EPEC.2012.6474970

[30]   B. Dunn, H. Kamath, and J.-M. Tarascon, "Electrical energy storage for the grid: A aattery of choices," *Science*, vol. 334, pp. 928–935, 2011.

[31]   A. Saez-de-Ibarra *et al.*, "Analysis and comparison of battery energy storage technologies for grid applications," in *2013 IEEE Grenoble Conference*, pp. 1–6, Jun. 2013, doi: 10.1109/PTC.2013.6652509

[32]   C. Spataru and P. Bouffaron, "30 – Off-grid energy storage," in *Storing Energy (2nd Edition)*, T. M. Letcher, Ed., Elsevier. 2022, pp. 731–752, doi: 10.1016/B978-0-12-824510-1.00033-7

[33]   K. C. Divya and J. Østergaard, "Battery energy storage technology for power systems—An overview," *Electr. Power Syst. Res.*, vol. 79, no. 4, pp. 511–520, Apr. 2009, doi: 10.1016/j.epsr.2008.09.017

[34]   C. S. Makola, P. F. Le Roux, and J. A. Jordaan, "Comparative analysis of lithium-ion and lead–acid as electrical energy storage systems in a Grid-Tied Microgrid application," *Appl. Sci.*, vol. 13, no. 5, p. 3137, Jan. 2023, doi: 10.3390/app13053137

[35]   W. H. Zhu, Y. Zhu, and B. J. Tatarchuk, "A simplified equivalent circuit model for simulation of Pb–acid batteries at load for energy storage application," *Energy Convers. Manag.*, vol. 52, no. 8, pp. 2794–2799, Aug. 2011, doi: 10.1016/j.enconman.2011.02013

[36]   D. Kumar, S. K. Rajouria, S. B. Kuhar, and D. K. Kanchan, "Progress and prospects of sodium-sulfur batteries: A review," *Solid State Ion.*, vol. 312, pp. 8–16, Dec. 2017, doi: 10.1016/j.ssi.2017.10.004

[37]   Y. Kim and N. Chang, "Background and related work," in *Design and Management of Energy-Efficient Hybrid Electrical Energy Storage Systems*, Y. Kim and N. Chang, Eds. Springer International Publishing, 2014, pp. 7–17, doi: 10.1007/978-3-319-07281-4_2

[38]   Y. Kim and N. Chang, *Design and Management of Energy-Efficient Hybrid Electrical Energy Storage Systems*. Springer International Publishing, 2014, doi: 10.1007/978-3-319-07281-4

[39]   Y. Gao, X. Zhang, Q. Cheng, B. Guo, and J. Yang, "Classification and review of the charging strategies for commercial lithium-ion batteries," *IEEE Access*, vol. 7, pp. 43511–43524, 2019, doi: 10.1109/ACCESS.2019.2906117

[40]   E. Banguero, A. Correcher, Á. Pérez-Navarro, F. Morant, and A. Aristizabal, "A review on battery charging and discharging control strategies: Application to renewable energy systems," *Energies*, vol. 11, no. 4, p. 1021, Apr. 2018, doi: 10.3390/en11041021

[41]   Y.-C. Chuang, Y.-L. Ke, H.-S. Chuang, and S.-Y. Chang, "Battery float charge technique using parallel- loaded resonant converter for discontinuous conduction operation," *IEEE Trans. Ind. Appl.*, vol. 48, no. 3, pp. 1070–1078, May 2012, doi: 10.1109/TIA.2012.2190961

[42]   A. C.-C. Hua and B. Z.-W. Syue, "Charge and discharge characteristics of lead acid battery and LiFePO4 battery," in *The 2010 International Power Electronics Conference, ECCE ASIA, Sapporo, Japan*, pp. 1478–1483, 2010, doi: 10.1109/IPEC.2010.5544506

[43]   X. Huan *et al.*, "A review of pulsed current technique for lithium-ion batteries, " *Energies*, vol. 13, no. 10, p. 2458, 2020.

[44]   R. F. Nelson, "Chapter 9 – Charging techniques for VRLA batteries," in *Valve-Regulated Lead-Acid Batteries*, D. A. J. Rand, J. Garche, P. T. Moseley, and C.

D. Parker, Eds. Elsevier, 2004, pp. 241–293, doi: 10.1016/B978-044450746-4/50011-8

[45] F. Mumtaz, N. Z. Yahaya, S. T. Meraj, B. Singh, R. Kannan, and O. Ibrahim, "Review on non-isolated DC-DC converters and their control techniques for renewable energy applications," *Ain Shams Eng. J.*, vol. 12, no. 4, pp. 3747–3763, 2021.

[46] Y. Koç, Y. Birbir, and H. Bodur, "Non-isolated high step-up DC/DC converters – An overview," *Alex. Eng. J.*, vol. 61, no. 2, pp. 1091–1132, Feb. 2022, doi: 10.1016/j.aej.2021.06.071

[47] A. Allehyani, "Analysis of a symmetrical multilevel DC-DC boost converter with ripple reduction structure for solar PV systems," *Alex. Eng. J.*, vol. 61, no. 9, pp. 7055–7065, Sep. 2022, doi: 10.1016/j.aej.2021.12.049

[48] C. H. Tran, F. Nollet, N. Essounbouli, and A. Hamzaoui, "Modeling and simulation of stand alone photovoltaic system using three level boost converter," in *2017 International Renewable and Sustainable Energy Conference (IRSEC)*, pp. 1–6, Dec. 2017, doi: 10.1109/IRSEC.2017.8477246

[49] M. A. Abundis, O. Carranza, J. J. Rodríguez, R. Ortega, and J. V. Chavez, "DC/DC converter for a PV system operating in stand-alone and grid-tied modes," in *2018 XXXI International Summer Meeting on Power and Industrial Applications (RVP-AI)*, pp. 64–69, Jul. 2018, doi: 10.1109/RVPAI.2018.8469781

[50] A. Elkhateb, N. A. Rahim, J. Selvaraj, and B. W. Williams, "DC-to-DC converter with low input current ripple for maximum photovoltaic power extraction," *IEEE Trans. Ind. Electron.*, vol. 62, no. 4, pp. 2246–2256, Apr. 2015, doi: 10.1109/TIE.2014.2383999

[51] Y. Zheng, B. Brown, W. Xie, S. Li, and K. Smedley, "High step-up DC–dc converter with zero voltage switching and low input current ripple," *IEEE Trans. Power Electron.*, vol. 35, no. 9, pp. 9416–9429, Sep. 2020, doi: 10.1109/TPEL.2020.2968613

[52] N. Aouchiche, M. S. Aitcheikh, M. Becherif, and M. A. Ebrahim, "AI-based global MPPT for partial shaded grid connected PV plant via MFO approach," *Sol. Energy*, vol. 171, pp. 593–603, Sep. 2018, doi: 10.1016/j.solener.2018.06.109

[53] S. Chtita, A. Derouich, S. Motahhir, and A. EL Ghzizal, "A new MPPT design using arithmetic optimization algorithm for PV energy storage systems operating under partial shading conditions," *Energy Convers. Manage.*, vol. 289, p. 117197, Aug. 2023, doi: 10.1016/j.enconman.2023.117197

[54] I. Saady, M. Karim, B. Bossoufi, N. El Ouanjli, S. Motahhir, and B. Majout, "Optimization and control of photovoltaic water pumping system using kalman filter based MPPT and multilevel inverter fed DTC-IM," *Results Eng.*, vol. 17, p. 100829, Mar. 2023, doi: 10.1016/j.rineng.2022.100829

[55] R. B. A. Koad, A. F. Zobaa, and A. El-Shahat, "A novel MPPT algorithm based on particle swarm optimization for photovoltaic systems," *IEEE Trans. Sustain. Energy*, vol. 8, no. 2, pp. 468–476, Apr. 2017, doi: 10.1109/TSTE.2016.2606421

[56] H. Rezk, M. M. Zaky, M. Alhaider, and M. A. Tolba, "Robust fractional MPPT-based moth-flame optimization algorithm for thermoelectric generation applications," *Energies*, vol. 15, no. 23, p. 8836, Jan. 2022, doi: 10.3390/en15238836

[57] K. Zeb *et al.*, "A comprehensive review on inverter topologies and control strategies for grid connected photovoltaic system," *Renew. Sustain. Energy Rev.*, vol. 94, pp. 1120–1141, Oct. 2018, doi: 10.1016/j.rser.2018.06.053

[58] X. Zhu, H. Wang, W. Zhang, H. Wang, X. Deng, and X. Yue, "A novel single-phase five-level transformer-less photovoltaic (PV) inverter," *CES Trans. Electr. Mach. Syst.*, vol. 4, no. 4, pp. 329–338, Dec. 2020, doi: 10.30941/ CESTEMS.2020.00040

[59] B. Chen and J.-S. Lai, "A family of single-phase transformerless inverters with asymmetric phase-legs," in *IEEE Applied Power Electronics Conference and Exposition (APEC)*, vol. 2015, pp. 2200–2205, 2015, doi: 10.1109/APEC.2015.7104654

[60] A. Stone, Md. Rasheduzzaman, and P. Fajri, "A review of single-phase single-stage DC/AC boost inverter topologies and their controllers," in *2018 IEEE Conference on Technologies for Sustainability (SusTech)*, pp. 1–8, Nov. 2018, doi: 10.1109/ SusTech.2018.8671380

[61] R. Duarte, Á. García-Riazuelo, L. A. Sáez, and C. Sarasa, "Analysing citizens' perceptions of renewable energies in rural areas: A case study on wind farms in Spain," *Energy Rep.*, vol. 8, pp. 12822–12831, Nov. 2022, doi: 10.1016/j.egyr.2022. 09.173

[62] C. Ammari, D. Belatrache, B. Touhami, and S. Makhloufi, "Sizing, optimization, control and energy management of hybrid renewable energy system—A review," *Energy Built Environ.*, vol. 3, no. 4, pp. 399–411, Oct. 2022, doi: 10.1016/ j.enbenv.2021.04.002

[63] R. Satpathy and V. Pamuru, "Chapter 7 – Off-grid solar photovoltaic systems," in *Solar PV Power*, R. Satpathy and V. Pamuru, Eds. Academic Press, 2021, pp. 267–315, doi: 10.1016/B978-0-12-817626- 9.00007-1

[64] J. Li, P. Liu, and Z. Li, "Optimal Design and Techno-Economic Analysis of a Solar-Wind-Biomass Off-Grid Hybrid Power System for Remote Rural Electrification: A Case Study of West China – ScienceDirect." *Energy*, Vol. 208, p. 118387. Oct 2020, doi: 10.1016/j.energy.2020.118387 (accessed May 09, 2023).

[65] S. Sinha and S. S. Chandel, "Review of software tools for hybrid renewable energy systems," *Renew. Sustain. Energy Rev.*, vol. 32, pp. 192–205, Apr. 2014, doi: 10.1016/ j.rser.2014.01.035

[66] "Welcome | TRNSYS: Transient System Simulation Tool." www.trnsys.com/ (accessed Jun. 07, 2023).

[67] N. R. Canada, "retscreen," Mar. 10, 2010. https://natural-resources.canada.ca/ maps-tools-and-publications/tools/modelling-tools/retscreen/7465 (accessed Jun. 08, 2023).

[68] "HOMER – Hybrid Renewable and Distributed Generation System Design Software." www.homerenergy.com/ (accessed Jun. 08, 2023).

[69] K. Ram, P. K. Swain, R. Vallabhaneni, and A. Kumar, "Critical assessment on application of software for designing hybrid energy systems," *Mater. Today Proc.*, vol. 49, pp. 425–432, Jan. 2022, doi: 10.1016/j.matpr.2021.02.452

[70] H. El-Houari *et al.*, "Design, simulation, and economic optimization of an off-grid photovoltaic system for rural electrification," *Energies*, vol. 12, no. 24, p. 4735, Jan. 2019, doi: 10.3390/en12244735

[71] S. R. Spea and H. A. Khattab, "Design sizing and performance analysis of stand-alone PV system using PVsyst Software for a location in Egypt," in *2019 21st International Middle East Power Systems Conference (MEPCON)*, pp. 927–932, Dec. 2019, doi: 10.1109/MEPCON47431.2019.9008058

[72] R. Kumar, C. S. Rajoria, A. Sharma, and S. Suhag, "Design and simulation of stand-alone solar PV system using PVsyst Software: A case study," *Mater. Today Proc.*, vol. 46, pp. 5322–5328, Jan. 2021, doi: 10.1016/j.matpr.2020.08.785

[73] I. Batarseh and K. Alluhaybi, "Emerging opportunities in distributed power electronics and battery integration: Setting the stage for an energy storage revolution," *IEEE Power Electron. Mag.*, vol. 7, no. 2, pp. 22–32, Jun. 2020, doi: 10.1109/MPEL.2020.2987114

8 Peer-to-Peer Transactions in the Battery-Integrated Residential Sector

Debasmita Panda and Altaf Q. H. Badar

8.1 INTRODUCTION

Rapid increase in renewable penetration along with smart grid technologies have drastically changed the way generation and consumption is taking place [1]. This change has added new grid participants as prosumers so as to facilitate the operation of a decentralized and open power market. Prosumers are both producers and consumers. This paradigm shift requires new storage technologies, communications changes (e.g., ICT and smart meters), new control algorithms (transactive control), computational methods, and advanced data sharing platforms (e.g., AMI and blockchain) to handle this distribution network peer-to-peer (P2P) energy market model.

With the ambition of reducing carbon emission and driven by regulatory incentives, DERs installation is increasing worldwide. The net generation from small-scale solar PV in the USA is reported to have increased from 11,233 GWh to 58,512 GWh between 2014 and 2022 [2]. The peer-to-peer market model makes it possible for the prosumers to trade energy in their local surroundings as well as to support the utility grid by providing ancillary services when needed. Various studies have been performed over the past few years on local energy market trading focusing on a range of aspects, namely the design of the market, its policy aspect, trading platforms, communication facilities, and societal impact [3, 4]. The P2P energy market provides opportunities for small producers equipped with solar PV panels, micro turbines, and battery energy storage systems (BESS) in creating a new electricity supply chain that can lead to effective utilization of small-scale energy generations. In a broader sense, the P2P concept opens up corridors for microgrid operators for easy energy sharing.

8.1.1 PROCESS OF ELECTRICITY TRADING

Electricity trading in real markets occurs through a variety of channels, depending on the specific market structure and regulatory framework in place. Here is an overview of how electricity trading occurs in a typical wholesale market, as mentioned in [5].

1. *Market participants submit offers and bids.* Generators, suppliers, and other market participants submit offers to sell electricity and bids to buy electricity for specific delivery periods.

DOI: 10.1201/9781003441236-8

2. *Market clearing process.* The market clearing process matches the supply and demand bids to determine the price and quantity of electricity that will be traded. This process considers factors such as transmission constraints, available capacity, and marginal costs.

3. *Contract settlement.* After the market clearing process, contracts are settled between buyers and sellers, and the actual delivery of electricity occurs. Settlements are usually done through financial contracts rather than physical delivery, which allows for greater flexibility and cost savings.

4. *Balancing and scheduling.* Grid operators and other market participants continuously monitor and adjust the flow of electricity to ensure that supply and demand are balanced in real time. This balancing and scheduling process is critical to maintaining the stability and reliability of the electricity grid.

8.1.2 MOTIVATION OF P2P ENERGY SHARING CONCEPT

In a smart grid era, every peer is either producer or consumer or both. As per the concept of P2P energy sharing, when a prosumer sells its excess to the grid (peer-to-grid, P2G), it gets the benefit according to the feed-in tariff (FIT) rates. Conventionally, the FIT rates are introduced by the regulators to encourage the renewable producer to sell their excess energy to the grid. In recent times, most countries' FIT rates have reduced to a much lesser value than the actual grid clearing price (GCP), and so the trading freedom for these small prosumers is almost negligible. Then comes the P2P energy trading paradigm, which encourages local transactions through a marketplace.

8.1.3 CONCEPT OF DISTRIBUTION LOCAL ENERGY MARKET (D-LEM)

D-LEM is a market model that enables prosumers to trade energy in their local area as well as to support the utility grid whenever needed. The D-LEM may function in a centralized platform forming a community-based market [6], and it may perform in a decentralized platform where peers can share the surplus and demand bilaterally (P2P trading) [7]. The advantages of P2P-based D-LEM is the flexibility between the peers over their trading process while availing demand at a lower price. In centralized community-based D-LEM trading, the social welfare is maximized to find the market clearing results. The P2P and P2G market structure without storage systems is shown in Figure 8.1.

Based on the energy sharing mechanism and pricing mechanism followed in a D-LEM for P2P energy transactions, the market platforms can be categorized as decentralized markets (mostly contract type), centralized markets (uniform clearing at D-LEM [8, 9]), or compound markets (a combination of the previous two).

In a decentralized market mechanism, the energy sharing can happen in two ways: in a coordinated manner [10, 11] or as direct energy sharing [12–15]. A third party exists in the D-LEM to facilitate energy sharing in a coordinated manner. The third party can be an aggregator or energy service provider. The pricing mechanism followed within the peer level is decided by the individual prosumers based on their energy status and production cost. However, while trading at a P2G level, the price to buy from the grid is at grid price, and the price to sell to the grid is at FIT. Therefore, the internal price in such a scenario is bounded by the grid price and the FIT. In the

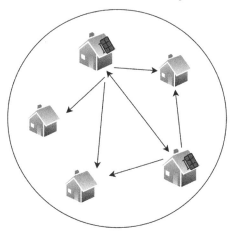

Decentralized Trading

FIGURE 8.1　Decentralized P2P market type.

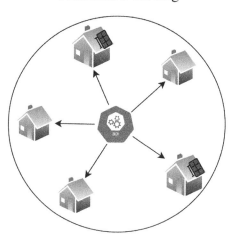

Centralized Trading

FIGURE 8.2　Centralized P2P market type.

direct energy sharing model, the users directly trade the energy without management by any third party, as shown in Figure 8.1, through bilateral negotiations.

In a centralized trading system, the aggregators communicate with the peers and manage the operation of DERs. The aggregator solves a centralized trading optimization with the aim of achieving maximum profit or maximizing social welfare, satisfying all operation constraints of the network to find a uniform clearing price, as shown in Figure 8.2. Proper selection of the centralized trading model is very important. The optimal trading strategy of peers can be formulated by framing an optimization problem with a set of constraints.

The compound or hybrid market type is a combination of decentralized and centralized arrangements where the peers can decide to switch from a conventional energy market to a P2P energy market with the price signal from the aggregator. The role of the aggregator here is to manage and supervise activities of the peers.

The objective function in any of centralized/hybrid market types can be defined as one of four types:

Type 1. A cost minimization problem, that is, the total cost of prosumer to trade in D-LEM. This cost includes the operating cost of an individual peer or multiple peers coming together through an aggregator forming a cost function.

Type 2. A profit maximization problem, that is, the profit incurred by selling energy in D-LEM. Such trading problems belong to producers and aggregators in the LEM.

Type 3. A social welfare maximization problem, that is, the societal benefit of the whole distribution network interns of the energy market is maximized by formulating a centralized optimization model.

Type 4. A game theory-based approach can be used to model the objective function when considering rivals' bidding strategy in a non-cooperative manner.

The objective function types 1 and 2 are the decentralized solution type, whereas type 3 provides a centralized solution. Type 4 provides a competitive solution where rival strategic bidding information is considered. The four objective functions get optimized satisfying system constraints as follows:

a. Power flow limits to ensure the lines connected between the peers are not overloaded beyond their capacity causing congestion in the network;

$$P_{flow}^{min}\left(t\right) \le P_{flow}\left(t\right) \le P_{flow}^{max}\left(t\right) \tag{8.1}$$

b. Limits on amount of power cleared for trading to ensure fair transactions by putting lower and upper bounds of total power traded by any particular trader I;

$$P_i^{min} \le P_i\left(t\right) \le P_i^{max} \tag{8.2}$$

c. System ramp up and ramp down constraints, if any; and
d. In case of storage unit consideration, the state of charge (SOC) of battery units, the minimum and maximum storage capacity, etc.

8.2 PEER-TO-PEER (P2P) TRANSACTIONS: MARKET STRUCTURE

The concept of P2P energy sharing with many prosumers in the distribution network has recently entered the business sector with technological advancements in the concepts of virtual power plant (VPP), demand-side management (DSM), BESS, and blockchain-enabled transactions. The P2P and P2G energy sharing structure without and with BESS is shown schematically in Figures 8.3 and 8.4, respectively. The concept of aggregate prosumers acting as an energy supplier thus forming a VPP can lead to the formation of LEM. This can help in integration of small residential prosumers to proactively participate in P2P/P2G energy sharing.

The concepts of VPP and community microgrid are two similar kinds of solutions in aggregation of renewable power generation, although they have specific differentiations in terms of physical structure and operation. Operationally, a microgrid can act both as stand-alone and grid-connected mode; a VPP can work only in grid-connected mode. Structurally, a microgrid is confined to a physical boundary, while a VPP can accommodate integration of a variety of sources with many consumers.

Emerging energy storage technologies in this area of P2P transactions can be listed as follows:

- Hydraulic pumped energy storage (HPES);
- Compressed air energy storage (CAES);
- Flywheel energy storage (FWES);
- Superconductor magnetic energy storage (SMES);
- Battery energy storage system (BESS);
- Super capacitor energy storage (SCES); and
- Hydrogen along with fuel cell (FC).

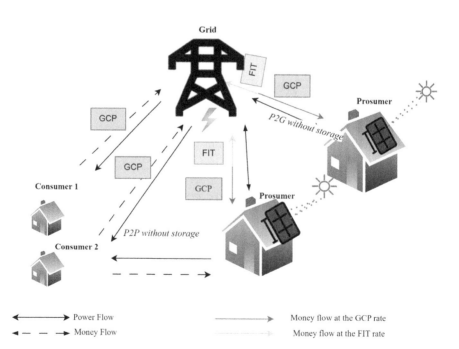

FIGURE 8.3   P2P energy sharing without storage system.

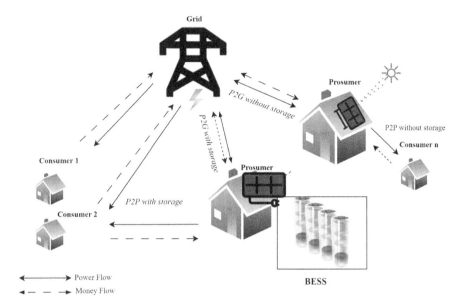

FIGURE 8.4 P2P/P2G energy sharing with storage system.

Of these technologies, BESS with lithium-ion batteries is the dominant player in the residential sector and small-scale applications, as it has nearly 100 percent energy storage efficiency and high energy density.

8.3 PRICING MECHANISM OF THE P2P MARKET

Deregulation can bring about competition in terms of price fixation. From the fundamental principle of economics, pricing of any product depends on the supply and demand status of the product. If the supply is greater than the demand, the price falls, and vice versa. This can simply be applied in a P2P market. In general, there are three prices/costs/charges associated with a large prosumer: a generation cost, an import cost, and an export price. In India, we use the concept of levelized cost of electricity from solar PV for calculating the generation cost, which has reduced over a period of time. The import cost is the GCP, as we discussed earlier. GCP is decided based on a double auction market clearing mechanism through a centralized optimization model. The export price from the prosumer to the grid is at the FIT rate set by the regulators. To encourage self-consumption, the export price is often kept low compared to the import price. Therefore, practically, the P2P energy transaction price varies between the FIT (lower range) and GCP (upper range). Keeping this fundamental point in mind, various pricing mechanisms have been discussed in the literature.

For the residential sector, the pricing decision is based on the suitability of the market structure and its financial sustainability. The fundamental supply to demand ratio (SDR) concept is used to set a price for residential prosumer [15]:

Condition 1. SDR < 1: Supply < Demand → Deficit condition: Price rises

Condition 2. SDR = 0: Supply = 0; No generation available from prosumer (depends on the main grid for LEM demand)

Condition 3. SDR = 1: Supply = Demand; LEM is self-sufficient to meet its demand from its own generation

Condition 4. SDR > 1: Supply > Demand → Surplus condition: Price decreases

When there is a surplus, export takes place at FIT, and in case of deficit, import takes place at GCP. But to facilitate P2P, export/import can happen between peers at a 'lamda' price, where FIT < lambda < GCP. In such a case, the concept of SDR can be used to finalize the lambda price.

Calculation of SDR and P2P Prices
The SDR value can be calculated at any instant of time by taking the total energy supply divided by the total energy demand of peers:

$$SDR = total\ energy\ supply\ /\ total\ energy\ demand,\ of\ a\ local\ area$$

The LEM aggregator collects the aggregated information about the peers regarding their selling and purchasing power. Pricing calculation is done for the four conditions mentioned earlier. Figure 8.5 shows the possible P2P pricing mechanisms. Here the LEM aggregator acts as a third-party facilitator to facilitate energy transactions between peers and the grid.

Condition 1: SDR < 1
For deficit situations within the LEM, there may be peers with surplus but not enough to meet the demand of deficit peers. Mathematically, such a situation will occur if we consider (S1+S3) < D2, where S1 and S3 are the surplus generation available from peers 1 and 3, respectively, and D2 is the demand needed for peer 2.

To meet the fundamental objective of a P2P energy market, these peers will trade among themselves initially until available S1 and S3 are completely utilized.

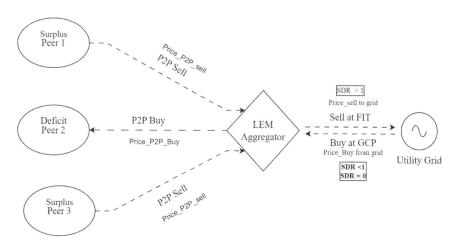

FIGURE 8.5 Illustration of possible pricing mechanism in a P2P energy market.

Remaining demand will be taken from the mail grid at the GCP rate. The price for such P2P trading is fixed by the LEM aggregator. The P2P sell price can be given by the following equation:

$$price_{p2P_{sell}} = \frac{price_{sell\,to\,Grid} \times price_{Buy\,from\,Grid}}{\left(price_{Buy\,from\,Grid} - price_{sell\,to\,Grid}\right) \times SDR + price_{sell\,to\,Grid}} \qquad (8.3)$$

The P2P buy price can be given as follows:

$$Price_{P2P_{Buy}} = Price_{P2P_{sell}} .SDR + price_{Buy\,from\,Grid}\left(1 - SDR\right) \qquad (8.4)$$

The role of aggregators here is to provide possible power balance in the LEM. They do so by recovering some amount of the service fee from peers to cover the power loss and other possible expenditures.

Condition 2: SDR = 0
SDR is 0 when no peer generation is available. In such a case, LEM total demand is met by the main grid. Such a condition will occur if we consider S1 = S3 = 0 & D2 > 0. The P2P pricing is given as follows:

$$price_{Buy\,from\,Grid} = GCP \qquad (8.5)$$

Condition 3: SDR = 1
LEM total generation from peers is equal to total LEM demand of all peers, which implies LEM is self-sufficient to meet its demand from its own available generation. Such a condition will occur when we consider S1+S3 = D2. The P2P buy and sell price can be determined by formulating a centralized optimization carried at the aggregator side. All the system constraints have to be satisfied during the optimization process for finding a uniform clearing price.

As shown in Figure 8.6, the bidding negotiation between the peers continues until both buyer and seller agree on a price and reach an equilibrium point, that is,

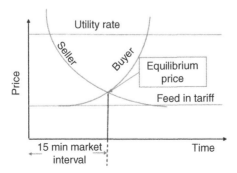

FIGURE 8.6   Centralized clearing mechanism.

the clearing price. This process needs to be completed within a specific time frame depending on the market timeline. As both buyers and sellers are involved in this bidding process, it is called a double auction mechanism.

Condition 4: SDR > 1

The LEM is in surplus, with total consumption less than its available generation and storage reserves from peers. Such a case results in selling the excess to the utility grid at FIT and will occur if we consider (S1+S3) > D2. The pricing for selling energy to the grid will be given by the following equation:

$$price_{sell\ to\ Grid} = \text{FIT} \tag{8.6}$$

8.4 TECHNOLOGIES FOR FACILITATING P2P ENERGY TRANSACTIONS IN THE RESIDENTIAL SECTOR

8.4.1 BLOCKCHAIN

Blockchain is a peer-to-peer distributed ledger [16, 17]. It is a distributed solution for recording asset transactions. The decentralized nature is used to verify every transaction via consensus and predefined validation mechanisms. Each block in a blockchain is time-stamped with a new block referring to the earlier block, which adds another feature to the blockchain, namely immutability. The emerging blockchain technology provides distributive and decentralized solutions for energy transactions [18]. Hence, there are several advantages of blockchain to both users and energy grid:

- It reduces overall economic cost;
- It provides security and immutability of data;
- It enhances transparency, as the ledger is distributed among all users;
- It facilitates decentralization of the energy supply;
- It reduces the computational burden; and
- It reduces the storage space required.

There are several blockchain platforms that are well suited for P2P energy trading. Listed below are some of the popularly used blockchain-based consensus algorithms with smart contracts:

- Ethereum is a popular Blockchain platform that offers smart contract functionality, making it well suited for P2P energy trading. It allows developers to create decentralized applications (dApps) that can execute transactions automatically based on pre-set conditions [19].
- Power Ledger is a blockchain-based platform specifically designed for P2P energy trading. It allows consumers to buy and sell renewable energy directly with each other, without the need for a middleman [20].
- Energy Web Chain is an open-source blockchain platform designed specifically for the energy sector. It is built on the Ethereum blockchain and offers a suite of tools for energy market participants, including P2P energy trading [21].

- WePower is a blockchain-based platform that enables P2P energy trading by allowing renewable energy producers to sell energy directly to consumers. It offers a range of tools for renewable energy project financing, energy accounting, and energy trading [22].
- Grid+ is a blockchain platform that offers a range of energy services, including P2P energy trading. It uses Ethereum-based smart contracts to enable secure, automated transactions between energy producers and consumers [23].

The cryptography feature in a blockchain helps in protecting information through the use of codes, so that only those for whom the information is intended can read and process it.

8.4.2 Peer-to-Peer Energy Transactions Using Blockchain

P2P energy transactions can be facilitated by blockchain in the following four steps.

Step 1: Peer Registration to Participate in the Trading Process
Any peer who wants to participate in the P2P trading process has to register through user-friendly **DApp** (**D**istributed **App**lication) applications. At a specific time a peer can act either as a buyer or as a supplier. There are a few standard distribution test systems available for carrying out studies related to P2P trading [20]. We have used the distribution bus system given in Figure 8.7 with N buses in the network and residential buildings placed at each node, where N is the total number of buses/nodes, M is the total number of prosumers connected a bus, Node 1 is the slack bus connected to the grid, and the surplus/demand at a particular bus = available generation at that bus – demand at that bus. At any particular instant, the surplus nodes registered as supplier and deficit nodes are registered as buyers through DApp to initiate the trading process.

Step 2: Selection and Design of Matching Strategy
Once the peer's registration is done, the buyers and sellers list is prepared and an appropriate selection and matching strategy is followed to clear the market with appropriate pricing structure. The various kinds of matching strategies available in the literature can be classified broadly into two categories [21]: the supply–demand matching strategy, and the distance-based matching strategy. In the supply–demand matching strategy, it is assumed that households with a surplus power are more likely

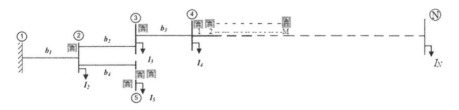

FIGURE 8.7   Radial distribution network with prosumers.

to trade energy with households with a power deficit, and vice versa. This strategy could be adopted to create a bilateral trading market where an economic mechanism is prioritized. The distance-based matching strategy is based on households' distance from each other in the network. The distance between households depends on the topology of the network, which is assumed to be known in this strategy. Generally, households are connected to distribution networks, which typically have a radial topology. This strategy is more focused on reducing long-distance electricity flows through the lines, potentially enhancing energy efficiency in the network by encouraging P2P trading among nearby households.

Step 3: Auction Model Considering Loss Cost
Once the matching strategy is decided, an auction model is formed to facilitate the trading. With the results from the matching strategy, an AC optimal power flow-based model is run satisfying all time constraints to check the voltage violation. Any voltage violation in the network is solved through an iterative process till convergence is achieved. Backward forward load flow method can be used to check the voltage violation limits, assuming that each bus is a residential peer/microgrid/prosumer that participates in the auction. The surplus and deficit status is found as follows:

$$\text{Surplus/demand} = \text{Generation at that bus} - \text{Demand at the bus}$$

For calculation of price quoted by the microgrid, we may use the levelized cost of energy (LCOE), a measure of the average net present cost of electricity generation for a generator over its lifetime. Let that price be represented as follows:

$$\text{Price}' = \text{Total cost over lifetime} / \text{Electrical energy produced over lifetime} \tag{8.7}$$

While trading in a P2P grid, the price of the seller bus is set to be summed up with the cost for line loss of energy the seller bus incurs while trading with the buyer bus. So the total trading price is as follows:

$$\text{Price} = \text{Price}' + \text{line loss cost} \tag{8.8}$$

Let i and j be the index of microgrids that are to participate in the energy trading at any time instant t and have a direct network connection with each other:

$$d_{ij,t} = \text{quantity of energy traded between microgrid } i \ \& \ j$$

Total traded energy between i (seller) and all of j (buyers) at time t is equal to net power available at that time:

$$P_{ij,t} = \sum_{j=1}^{M} d_{ij,t} \tag{8.9}$$

where M is the total number of buyers. We calculate the power loss in the line as follows:

$$Loss_{ij,t} = I_{ij,t}^2 R_{ij} \qquad (8.10)$$

$$Total\,loss\,cost = Penalty\,for\,loss\left(\frac{Rs}{kWh}\right) * Loss_{ij,t} \qquad (8.11)$$

At the end of this auction model, the peers would be able to get the trading results with individual trading price information. Once the trading results are published, using smart contracts in the blockchain the financial settlement will be made.

Step 4: Adding Features of Blockchain with a Smart Contract

Smart contracts are essential for Ethereum transactions to happen. Smart contracts are the fundamental building blocks that allow the conventional contacts into digital contracts. The smart contracts are written in the language called *solidity*, and the resulting smart contract can be used to initiate and end the digital transactions between users with their concern. After the Step 3 auction results, consider 'Party One' and 'Party Two' engaged in a P2P transaction. An illustration of smart contract execution between Party One and Party Two written in *solidity* (version 0.8.0) is shown in Figure 8.8.

The smart contract in Figure 8.8 defines a contract called 'Ether Transfer' that takes in two Ethereum addresses, Party One and Party Two, as well as a 'unit' value representing the amount of Ether to be transferred. The contract also includes two bool variables to track whether each party has accepted the transaction ('unit' and 'bool' are the data types in *solidity*; 'unit' stands for unified integers, and 'bool' stands for Boolean).

```solidity
pragma solidity ^0.8.0;

contract EtherTransfer {
    address payable public partyOne;
    address payable public partyTwo;
    uint public amount;
    bool public acceptedByPartyOne;
    bool public acceptedByPartyTwo;

    constructor(address payable _partyOne, address payable _partyTwo, uint _amount) {    ⎫  A constructor to initialize
        partyOne = _partyOne;                                                            ⎬  both the users
        partyTwo = _partyTwo;
        amount = _amount;}                                                               ⎭

    function accept() public {
        require(msg.sender == partyOne || msg.sender == partyTwo, "Only party one or party two can accept");    ⎫
        if (msg.sender == partyOne) {                                                    ⎪  A function to accept the
            acceptedByPartyOne = true;                                                   ⎬  ether sent by one party to
        } else {                                                                         ⎪  other.
            acceptedByPartyTwo = true;
        }
        if (acceptedByPartyOne && acceptedByPartyTwo) {
            transfer();
        }}                                                                               ⎭

    function transfer() private {                                                        ⎫  A function to initiate
        require(address(this).balance >= amount, "Insufficient funds");                  ⎬  transfer of ether from one
        partyTwo.transfer(amount);   }                                                   ⎭  party to other

    function cancel() public {
        require(msg.sender == partyOne, "Only party one can cancel");                     ⎫  A function to cancel the
        selfdestruct(partyOne);   }                                                      ⎬  transaction.
    receive() external payable {}                                                        ⎭
}
```

FIGURE 8.8 Illustration of a smart contract written in *solidity*.

- The accept function can be called by either party to indicate that they accept the transaction. If both parties have accepted, the transfer function is called to transfer the Ether from PartyOne to PartyTwo.
- The cancel function can be called by PartyOne to cancel the transaction and return the Ether to PartyOne.
- The contract also includes a receive function to allow the contract to receive Ether.

Once the Ether transaction is done, a block is created with the trading results and the financial settlement done between the peer parties. The created block is then shared between the peers to maintain transparency and form a distributed ledger.

8.4.3 Case Study Showing the Advantages of P2P Energy Sharing

In the N bus system shown in Figure 8.7, let there be a load at bus B2, and PV generation available at buses 1 (B1) and 3 (B3), respectively. The 24h load data for B2, and the generation data of B1 and B3 are shown in the graphs below. After B1, B2, and B3 register as peers with their demand and generation information, using distance-based matching strategy, the buyers and sellers list is prepared.

The 24h load profile and PV generation profiles considered for this study are given in Figure 8.9. By solving the auction model developed in Step 3, we can observe the cost saving in (Rs. $\times 10^3$) with P2P trading compared to trading without P2P (solely depending on the main grid), as shown in Figure 8.10.

The blue line in the graph depicts the cost of energy if it is acquired solely from the grid, whereas the orange line depicts the cost of energy when there is supply from the energy trading market as well as from the grid. As depicted in the graph, during the day, when solar generation is high, with P2P transactions the cost saving to serve the same load is significant. This indicates that the P2P energy trading market is essential in the modern day, as it encourages the use of renewable resources and also significantly decreases the rate at which energy is brought by decentralized energy sources.

8.4.4 Facilitating Transactive Energy Trading in a Smart Grid

The blockchain model discussed above is a three-layered architecture consisting of data layers, a physical layer, and a blockchain layer, as shown in Figure 8.11.

The data layer will handle the forecasting of DER generations and consumer demand patterns. This information is then forwarded to the physical layer, where the grid operator will perform market clearing and will send the information to the blockchain layer. The stepwise process (mentioned in Figure 8.11) followed in the blockchain-based energy trading platform is as follows:

- In steps 1 and 2, all the trading participants (namely, prosumer peer, consumer, and T-DSO) will be registered in the Ethereum blockchain and their trading e-wallet will be created by linking public and private keys.

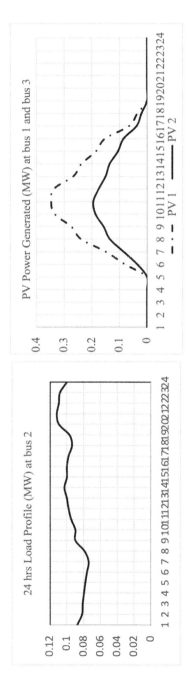

FIGURE 8.9   24h load data for bus 2, and generation data of bus 1 and bus 3 (MW).

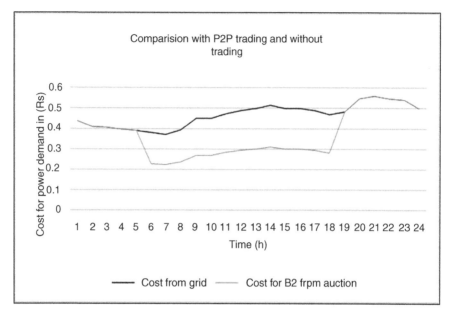

FIGURE 8.10   Comparison with and without P2P trading in terms of cost of energy.

- Steps 3 and 4 aim to list a prosumer as a buyer/supplier. The blockchain adminis-
  trator in the proposed trading system will verify the request and publish it in the
  blockchain. Once all the seller information is available, the same can be accessible to all potential buyers. Here the buyer can be a prosumer microgrid (consumer). Accordingly, all the buyer information is also recorded in the blockchain.
- In steps 5 and 6, the trading result is published after solving the trading optimization.

All these trading steps will be executed using a smart contract. A smart contract will
list the trading participant information, energy exchange, settlement after trading, and
payment functions. The smart contract captures different trading scenarios: between-
peers trading, peer-to-operator trading, and operator-to-peer trading.

- The trading results are then published and shared with all participants in the
  blockchain, completing steps 7 and 8. The crypto-currency coin Ether is then
  shared between the cleared participants.

Along with blockchain, artificial intelligence and machine learning (AI/ML)
techniques and the Internet of Things (IOT) help in the trading process by providing
robust solutions and advanced communication.

8.4.5 ARTIFICIAL INTELLIGENCE AND MACHINE LEARNING

Machine learning techniques have become increasingly popular in the field of
electrical load demand and renewable generation forecasting. These techniques

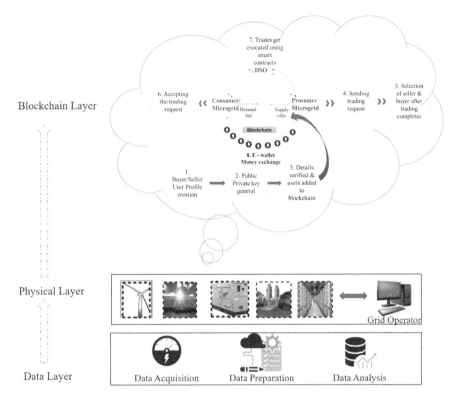

FIGURE 8.11   Blockchain-enabled transactive energy trading in smart grid energy market.

use historical data to predict future load demand, which can help utilities and grid operators better manage their systems and plan for future needs. One of the most popular machine learning techniques for load forecasting is neural networks. Neural networks are modeled after the structure of the human brain and consist of layers of interconnected nodes. These networks are trained using historical load data and can then be used to predict future demand. One advantage of neural networks is their ability to identify patterns and relationships within data that may be difficult for humans to recognize. The support vector machine, random forest, and a combination of these provide hybrid approaches to handle large and complex datasets, identify patterns and relationships within the data, and create accurate predictions of renewables and future load demands.

Again in the area of forecasting, researchers concentrated on predicting load demand with either single time-stamp ahead prediction (STAP) or multiple time-stamp ahead prediction (MTAP) [25]. In the case of STAP, the testing dataset is prepared the same as the training dataset using the actual values and predicting the next successive time-stamp demand value, here we require the actual future values in the test data to predict the load demand from the second time-stamp onwards. In the case of MTAP, the testing dataset is prepared the same as the training dataset, where the target vector in the sample is considered as multiple time-stamps up to which one is desired to forecast, here the testing dataset consists of only one sample to predict. The accuracy of P2P trading results depends on a suitable forecasting technique that can be used for accurately forecasting the uncertain renewable generation and load demand which may further help in near real-time P2P market operation.

8.4.6 INTERNET OF THINGS (IoT)

IOT monitoring and energy management is achieved using robotized electronic equipment [26]. Smart meters with IOT-based technologies may help in achieving demand-side management [27] in a P2P market. The non-essential and essential loads at residential buildings can be controlled using smart IOT-enabled devices. Technologies like blockchain, AI/ML, and IOT are surely going to revolutionize the distribution-level local energy market.

8.5 CONCLUSIONS

Blockchain-based peer-to-peer (P2P) energy trading is a rapidly developing field that promises to revolutionize the energy industry. Here are some potential future trends in this area:

- *Increased adoption of P2P energy trading platforms.* As more people become aware of the benefits of P2P energy trading, the adoption of blockchain-based platforms is likely to increase. This could lead to a more decentralized energy system where individuals and businesses can trade energy directly with each other.
- *Integration with renewable energy sources.* P2P energy trading could become an important tool for managing the intermittent output of renewable energy sources such as solar and wind. By allowing users to trade energy directly, P2P platforms could help to balance supply and demand in real time.
- *Emergence of microgrids.* A microgrid is a small-scale power grid that can operate independently of the larger grid. P2P energy trading could be used to facilitate the creation of microgrids, where neighbors or businesses could trade energy with each other within a local area.
- *Integration with electric vehicle charging.* P2P energy trading could be used to facilitate the charging of electric vehicles, allowing vehicle owners to buy and sell energy directly with each other.

Overall, blockchain-based P2P energy trading has the potential to transform the energy industry by creating a more decentralized and efficient system. As the technology continues to evolve, we can expect to see even more innovative applications in the years to come.

REFERENCES

[1]   Y. Luo and P. Davis, "Autonomous cooperative energy trading between prosumers for microgrid systems", *3rd IEEE international workshop on global trends in smart cities*, goSMART, 2014.

[2]   Electricity Data - U.S. Energy Information Administration (EIA). (n.d.). https://www.eia.gov/electricity/data.php

[3]   S. Zhou, et al., "A smart community energy management scheme considering user dominated demand side response and P2P trading", *Int. J. Electr. Power Energy Syst.*, vol. 114, p. 105378, 2020.

[4]   C. Long, Y. Zhou, and J. Wu, "A game theoretic approach for peer to peer energy trading", *Energy Proc.*, vol. 159, pp. 454–459, 2019.

[5]   Juhar Abdella, Zahir Tari, Adnan Anwar, Abdun Mahmood, and Fengling Han, "An architecture and performance evaluation of blockchain-based peer-to-peer energy trading", *IEEE Trans. Smart Grid*, vol. 12, no. 4, Jul. 2021, pp. 3364–3378.

[6]   Y. Parag and B. K. Sovacool, "Electricity market design for the prosumer era", *Nat. Energy*, vol. 1, no. 4, pp. 1–6, 2016.

[7]   Y. Zhou, J. Wu, and C. Long, "Evaluation of peer-to-peer energy sharing mechanismsbased on a multiagent simulation framework", *Appl. Energy*, vol. 222, pp. 993–1022, 2018.

[8]   C. Long, J. Wu, Y. Zhou, and N. Jenkins, "Peer-to-peer energy sharing through a two-stage aggregated battery control in a community Microgrid", *Appl. Energy*, vol. 226, pp. 261–276, 2018.

[9]   M. R. Alam, M. St-Hilaire, and T. Kunz, "An optimal P2P energy trading model for smart homes in the smart grid", *Energy Effic.*, vol. 10, no. 6, pp. 1475–1493, 2017.

[10]   Z. Zhang, R. Li, and F. Li, "A novel peer-to-peer local electricity market for joint trading of energy and uncertainty", *IEEE Trans. Smart Grid*, vol. 11, Mar. 2020, pp. 1205–1215.

[11]   T. AlSkaif, J. L. Crespo-Vazquez, M. Sekuloski, G. van Leeuwen, and J. P. S. Catalão, "Blockchain-based fully peer-to-peer energy trading strategies for residential energy systems", *IEEE Trans. Ind. Inf.*, vol. 18, Jan. 2022, pp. 231–241.

[12]   T. Morstyn, A. Teytelboym, and M. D. McCulloch, "Bilateral contract networks for peer-to-peer energy trading," *IEEE Trans. Smart Grid*, vol. 10, no. 2, pp. 2026–2035, Mar. 2019.

[13]   M. Andoni, V. Robu, D. Flynn, S. Abram, D. Geach, D. Jenkins, P. McCallum, and A. Peacock, "Blockchain technology in the energy sector: A systematic review of challenges and opportunities," *Renew.Sustain. Energy Rev.*, vol. 100, pp. 143–174, Feb. 2019.

[14]   Z. Lia, S. Bahramiradb, A. Paasob, M. Yana, and M. Shahidehpoura, "Blockchain for decentralized transactive energy management system in networked microgrids", *Elect. J.,* vol. 32, 06 Apr. 2019, pp. 58–72.

[15]   M. R. Hamouda, M. E. Nassar, and M. M. A. Salama, "A novel energy trading framework using adapted Blockchain technology," *IEEE Trans. Smart Grid*, vol. 12, no. 3, pp. 2165–2175, May 2021.

[16]   N. Liu, X. Yu, C. Wang, C. Li, L. Ma and J. Lei, "Energy-sharing model with price-based demand response for microgrids of peer-to-peer prosumers," *IEEE Trans. Power Syst.*, vol. 32, no. 5, pp. 3569–3583, Sept. 2017, doi: 10.1109/TPWRS.2017.2649558.

[17]   M. R. Hamouda, M. E. Nassar, and M. M. A. Salama, "Centralized blockchain-based energy trading platform for interconnected microgrids", *IEEE Access*, vol. 9, Jun. 17, 2021, pp. 95539–95550.

[18]   Z. Dong, F. Luo, and G. Liang, "Blockchain: A secure, decentralized, trusted cyber infrastructure solution for future energy systems," *J. Modern Power Syst. Clean Energy*, vol. 6, no. 5, pp. 958–967, 2018.

[19]   Ethereum.org. (n.d.). Home. ethereum.org. https://ethereum.org/en/

[20]   Powerledger. (n.d.). http://www.powerledger.io/

[21]　Energy web. Energy Web. (n.d.). http://www.energyweb.org/

[22]　WePower. (n.d.). http://www.blockdata.tech/profiles/wepower

[23]　GridPlus. (n.d.). https://gridplus.io/

[24]　B. Harish, D. Panda, K. R. Konda and A. Soni, "A comparative study of forecasting problems on electrical load time series data using deep learning techniques", *2023 IEEE/IAS 59th Industrial and Commercial Power Systems Technical Conference (I&CPS)*, Las Vegas, NV, USA, 2023, pp. 1–5, doi: 10.1109/ICPS57144.2023.10142125.

[25]　A. Burgio, D. Cimmino, A. Nappo, L. Smarrazzo, and G. Donatiello, "An IoT-based solution for monitoring and controlling battery energy storage systems at residential and commercial levels", *Energies,* vol. 16, no. 7, 2023, p. 3140.

[26]　B. E. Sedhom, M. M. El-Saadawi, M. S. El Moursi, M. A. Hassan, A. A. Eladl, "IoT-based optimal demand side management and control scheme for smart microgrid," *International Journal of Electrical Power & Energy Systems*, vol. 127, 2021, 106674, ISSN 0142-0615.

9 Benefits of Decentralized Residential Batteries for System-Level Energy Management

Shubham Yadav, Bhavna Jangid,
Chandra Prakash, Ajay Verma, Arun Nayak,
Rohini Haridas, Swasti Swadha, Shivani Garg,
Parul Mathuria, and Rohit Bhakar

9.1 INTRODUCTION

Greenhouse gas (GHG) emissions have increased the Earth's average temperature by 1.1 °C as compared to the 1800s. The United Nations Intergovernmental Panel on Climate Change has advised all nations to reduce emissions by 45 percent by 2030 and stop this rise in temperature at 1.5 °C to reach net zero targets [1]. The key to achieving this target is the integration of renewable energy sources (RES) into the supply and demand sides of the current power system architecture. It is resulting in the global and uneven penetration of distributed energy resources (DER) at the distribution level. They are located behind the meter (BTM) in the distribution system [2, 3]. Most of the BTM-DERs are in the form of rooftop solar PV and EVs, thus changing the structure of the distribution system. The grid dependency on DER penetration can create grid imbalances, increase demand for infrastructure expansion, and lead to transmission line losses [4, 5]. Additionally, this penetration of DERs with time generates the need to find storage solutions at the highest priority [5]. More grid flexibility is needed because of DERs' volatile generation, non-correlated supply–demand peaks, and voltage fluctuations caused by supply–demand mismatches [6–8]. Energy storage solutions (ESS) have emerged as a way to mitigate fluctuations caused by the inclusion of DERs, provide grid flexibility, and increase the profit margin for prosumers, utilities, and system operators [4, 7–10]. For ESS to be deployed, consumers of electricity must be identified.

Consumers of electricity can be broadly categorized into the following categories: residential, commercial, industrial, and transportation. Some electricity demand is of a must-serve nature, for instance, in residential, hospital, and industrial settings [11]. Among all these consumers, residential consumers' share of the consumption of electrical energy is significant. They accounted for 38 percent of total electricity consumption in the UK for the year 2021 [12], 38.9 percent in the USA for the year 2022 [13], 27 percent in the EU for the year 2021 [14], and domestic and agricultural consumers accounted for 26 percent and 18 percent, respectively, for the year 2021 [15]. Residential consumers are usually only concerned with uninterrupted supply along with economic investments. In order to fulfill needs and save money, they were motivated to store energy when excess power was available or electricity prices were low, which led to the installation of batteries in their households. The individual battery owners did not coordinate among themselves or with the grid. This implies that battery integration behind the meter has existed in an uncoordinated and decentralized manner since the very beginning. When batteries are connected in the distribution system at the individual (household, building) or community level, installation, control, operation, and maintenance lie in the hands of the owner, which is the consumer itself, and thus the batteries are referred to as decentralized batteries. This storage falls under small-scale energy storage and is deployed at BTM.

Rapid growth and adoption of rooftop solar PV are evidence to determine both the collaborative and non-cooperative power of consumers as a whole [16]. This necessitates involving them in providing system-level services, apart from using storage to fulfill their basic needs. It motivates devising methods or ways to protect the residential sector and thus benefit the power system. Their contribution can be encouraged by providing them with benefits in the form of incentives or programs. Policies supporting BESS-BTM, along with DER, can drive their actions towards a cleaner future.

This chapter is an effort to explain the role of decentralized residential ESS in enhancing the integration of RES [17]. It supports system-level energy management, benefiting residential consumers, utilities, and system operators. It will clarify and identify the reasons to emphasize the benefits of decentralized residential batteries to serve system-level energy management. Section 9.2 elaborates on ESS, followed by BESS, grid-scale battery versus residential batteries, the selection criteria for choosing a particular battery type, and BTM. Section 9.3 examines energy management systems and the inclusion of ICT in power systems. Section 9.4 details energy management through residential involvement along with the financial aspects to be considered during battery adoption. Section 9.5 covers the benefits of decentralized residential batteries for system-level energy management and the significance of cost–benefit analysis. Section 9.6 contains the case studies supporting this chapter. Section 9.7 addresses the challenges that can persist even after the deployment of BTM-ESS. Section 9.8 describes the government's initiatives in India to support BESS. Section 9.9 gives the conclusion of the chapter.

9.2   ELECTRICAL ENERGY STORAGE SOLUTIONS

ESS can solve many issues for the existing power system. At the grid level, it can be useful in reducing losses in the transmission and distribution networks through proper management of the local use of stored energy (that is, through decentralized batteries), or it can add to the flexibility between demand and generation [6, 18]. Storage potentially helps residential consumers save on their electricity bills. Different ways to save money are better explained in the later sections.

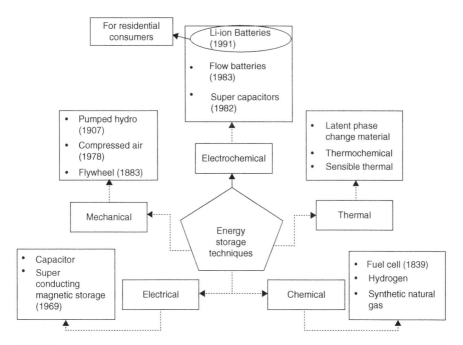

FIGURE 9.1   Types of energy storage techniques.

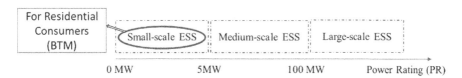

FIGURE 9.2   Small-scale, medium-scale, and large-scale ESSs.

Storage helps all the participants in the power system, including end consumers, utilities, and system operators, to save money or earn profits. Different electrical ESS have been invented since as long ago as 1839 [19]. Electrical, thermal, mechanical, chemical, and electrochemical techniques all have the potential to store electrical energy [19–22]. The types of storage and the year of their invention are shown in the schematic diagram in Figure 9.1.

Broadly, ESSs can be classified based on their power rating (PR), as illustrated in Figure 9.2. As small-scale ESSs are connected closer to the consumer site through the distribution system, they are known as BTM-ESS. For residential consumers, the rating is up to 5kW/13.5 kWh, and for commercial and industrial use, it is 5MW/10 MWh [23]. High costs continue to be a hindrance to adoption, yet batteries tend to be an accessible option for the BTM-ESS. Discussions have shown that residential consumers can adopt batteries based on their preferences. Hence, it is necessary to dive deep into battery energy storage solutions and discuss the selection criteria that should be considered when choosing a certain battery type. This section is further subdivided into BESS, selection criteria for batteries, and the BTM.

9.2.1 BATTERY ENERGY STORAGE SOLUTIONS

By far the majority of the world's energy storage capacity is still derived from pumped-storage hydropower. The 2022 hydropower status report by the International Hydropower Association (IHA) concluded that even with this rate of increasing hydropower storage deployment, net zero targets are a far-sighted goal to achieve. Only China has kept up with the pace on the net zero pathway for hydropower [24]. BESS thus acts as a companion to enhance this pace and help to create a cleaner future. Batteries are one of the electrochemical techniques to store electrical energy, which constitutes the BESS.

Batteries used by residential consumers are mainly of two types: static batteries in households and mobile energy storage systems in the form of EVs. EVs can use vehicle-to-grid (V2G) technology for charging and discharging in two directions on the grid. V2G enables the flow of energy, money, and information between the EV owner and the grid to balance supply and demand. This technology provides vehicles with the ability to recharge power stored from the electrical grid in a local power distribution system. It is particularly helpful in situations where there is a grid outage due to natural disasters or other unanticipated events [25, 26]. Some other technologies can be such that electrical energy is exported from V2G: EV to grid, V2H: EV to home or business, and V2L: EV to loads/other EV [27].

The benefits of installing BESS can be quantified by parameters such as payback period (PBP) and metrics like internal rate of return (IRR), and net present value (NPV) [28]. Residential batteries provide a set of benefits to the power system aside from fulfilling individual household needs. Batteries thus chosen for BESS should be selected wisely. The criteria that should be considered are discussed in the upcoming section.

The International Energy Agency (IEA) reports that between 2015 and 2020, the installed capacity of grid-scale BESS increased five times globally [6]. BESS can be employed at the grid scale, which is in front of the meter, or at the residential scale. Table 9.1 lists the differences between grid-scale and residential batteries. These differ on many grounds, including power capacity, location, purpose, cost and complexity, and ownership and control, which are discussed in this chapter.

Selection Criteria for Batteries
The selection of an appropriate battery can play an important role in BESS. The selection can be based on technical parameters, economics, and maturity [20]. Technical parameters include cycle life, rated power capacity, energy capacity, state of charge, self-discharge, and round-trip efficiency [20, 29].

- *Cycle life* refers to the number of cycles (from full charge to full discharge) a BESS provides before failure or breakdown.
- *Rated power capacity* (kW/MW) is the maximum rate of discharge a battery can achieve.

TABLE 9.1
Differences between grid-scale and residential batteries

Criterion	Grid-scale battery	Residential battery
Power capacity	Typically in the MW and GW range	Typically in the kW range
Location	Centralized locations (substations or power plants)	Decentralized locations (private properties, households)
Purpose	Grid stability	Backup power during blackouts
	Energy reliability	Reduce electricity costs through load
	Adoption of green energy sources	shifting and demand charge reduction
		Increase self-consumption
	Peak shaving	
	Frequency regulation	
	Backup power	
Cost	More expensive due to size and power capacity	Less expensive size and power capacity
Ownership and control	Utilities	Individual homeowners
	Independent power providers	Buildings or communities
	Grid operators	

- *Energy capacity* (kWh/MWh) is the maximum amount of energy that can be stored.
- *State of charge (SOC)* refers to the ratio of available capacity to the maximum charge that can be stored in the battery, and it is expressed as a percentage.
- *Self-discharge* is a phenomenon where a battery gets discharged on its own without supplying energy to the grid or consumer. It happens due to chemical reactions within the battery. It is expressed as the ratio of the total capacity of the battery to the number of days it can be stored without usage. It is an important parameter to be considered for long-term battery applications.
- *Round-trip efficiency* is the ratio of energy put into the battery to the energy retrieved from the battery, and it is also expressed as a percentage. It signifies the total AC-/AC or DC-DC efficiency of the BESS; losses due to self-discharge are included.

Economic and societal criteria include investment and operating costs, geographical location, and losses indicated by transportation, performance, and environmental constraints [20, 30]. Residential consumers have access to a wide variety of residential battery types, including lithium-ion (Li-ion), lead-acid, nickel-cadmium (Ni-Cd), nickel-metal hydride (NiMH), alkaline, zinc carbon, and many more. Depending on the factors of intended application, cost considerations, performance requirements, and environmental impacts, the consumer chooses a battery.

These days, Li-ion batteries dominate the ESS industry [23]. Li-ion batteries have recorded a significant decrease in their cost in the past few years, and for the next ten years this pattern is predicted to continue. The reason behind this prediction is the increasing demand for EVs in the current market. Economies of scale and technological advancements in the performance of batteries are drivers of this situation. The total cost of battery operation is not only dependent on investment costs but also on other factors. They are evaluated through the levelized cost of storage (LCOS). These factors include operation and maintenance costs, along with end-of-life costs and the cost of electricity for charging [30]. The number of technical parameters also affects the cost of batteries; they have already been discussed in this section.

The technical parameters and economics of batteries have led consumers to choose Li-ion batteries for residential household purposes. They have a longer cycle life, a higher energy density, fast charging and discharging capacities, and lower self-discharge, scalability, safety, compatibility, and availability than other rechargeable batteries. Recent applications of Li-ion batteries have highlighted the significance of regulation and policies in supporting the implementation of technologies and improving their maturity.

Batteries, once selected, need to be linked with DER in such a way that it facilitates system-level energy management. Small-scale ESS at the consumer end along with a distributed generation (DG) form a system known as BTM.

9.2.2 BEHIND THE METER

A renewable energy system known as BTM is installed in a single structure or at several locations that are operated and managed by one entity privately, such as campuses or households. It is operated through DG, or DER and storage units, to meet all or part of the energy demand of the end user [31]. Since decision-making is not interfered with by public entities, the system functions in a decentralized and uncoordinated manner most of the time. The decisions regarding deployment as well as operation of DERs are being taken "behind the meter" exclusively. Consumers consider different costs, like equipment costs, electricity rates, finance, and subsidies available to them, while going for BESS. There is no involvement of utilities or planners in deciding upon the method and location of deployment of DER or BESS [23, 32].

One kind of DG source is photovoltaic systems. They function by taking in photons from the sun and using solar cells to turn them straight into direct current, or electricity. In order to improve the voltage and/or current output, several solar cells are connected to make PV modules. These modules may be placed in arrays to feed power into the grid or provide consumers with electricity [23]. The installed capacity for distributed photovoltaic systems has increased significantly worldwide during the past decade. Supporting monetary policies and a gradual decline in the price of PV systems have been the main drivers of this rise. Nowadays, a sizable share of the global renewable energy capacity is made up of PV generators. The total installed capacity of solar generation in the world is about 1053 GW [33], out of 3371.8 GW, in 2022 [34].

In earlier times, only utilities installed large-scale stationary ESS. With time, smaller ESSs have now come into play because of a gradual reduction in the costs involved and the introduction of self-generation due to DERs. The primary value proposition that drove the major adoption of BTM-ESSs was the capacity to raise the portion of the total RES production that is utilized directly by onsite demand, also known as self-consumption by prosumers, and enhance the resilience of the power supply for utilities [35]. Li-ion batteries are dominant in the market for small-scale ESS. Some other batteries, like flow batteries, lead-acid batteries, and ultracapacitors, are also installed and used to some extent. Consumers are not restricted to consumption only; they can self-generate and self-consume. These are installed at the end user's site in the distribution system to provide onsite services. This storage system is capable of fulfilling consumers' needs and has the potential to support the overall electricity network. It provides system-level benefits starting from the consumer end. The flow of benefits passes to system operators and then to the utilities. Figure 9.3 is a schematic diagram of the BTM system. The diagram clearly illustrates how households with a combination of rooftop solar PV (DER) and a BESS can manage self-generation, import from the grid, and export to the grid simultaneously.

The rooftop solar PV generates electricity, which can be utilized onsite, used to charge the BESS, and exported to the grid when excess generation is there. The household energy needs can now be fulfilled in two ways: one from generation through PV and the other from the grid itself. Similarly, BESS can be charged by onsite generation or from the grid. BESS can also export to the grid once it gets fully charged. The growth of DERs has given rise to an explosive increment in the installed capacity of the BTM-ESS.

Globally, countries such as Germany have rooftop solar PV panels installed along with BTM storage for more than 40 percent of the total installed rooftops [23], and Australia has set a goal to deploy 1 million storage devices under small-scale ESS and residential units by 2025 [36].

Earlier, energy flow was from the source (i.e., the generator) to the load(s). Prosumerism has enabled consumers to generate electricity on their own; thus, the direction of energy flow is now bidirectional. The introduction and usage of multiple

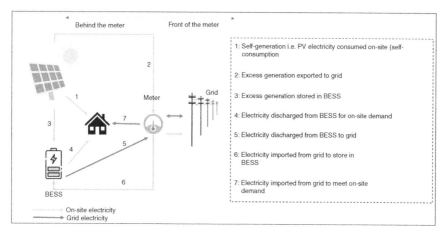

FIGURE 9.3   Schematic diagram of the BTM energy storage system.

energy resources, along with storage systems, formed the basis of the need for controlling the energy flow among these sources. The management of energy is needed to maximize the potential utilization of energy sources and storage systems involved. The process of easing energy management led to the application of information and communication technology (ICT) in the power system. Thus, the major changes that have occurred due to the introduction of ICT via smart meters are briefly covered.

9.3 ENERGY MANAGEMENT SYSTEMS AND ICT IN POWER SYSTEMS

Energy management is defined in the following terms: "Energy management applies to resources as well as to the supply, conversion, and utilization of energy. Essentially, it involves monitoring, measuring, recording, analyzing, critically examining, controlling, and redirecting energy and material flows through systems so that least power is expended to achieve worthwhile aims"; or "Energy management is the judicious and effective use of energy to maximize profits and enhance competitive positions through organizational measures and optimization of energy efficiency in the process" [37]. Existing literature suggests that this concept led to the development of different energy management systems. The energy management system (EMS) is defined as "a computer system comprising a software platform providing essential support services and a set of applications providing the functionality needed for the effective operation of electrical generation and transmission facilities to assure adequate security of energy supply at minimum cost" by International Electrotechnical Commission standard (IEC) 61970 [38]. EMS can be established in different environments, such as grid-connected, stand-alone, and microgrid in grid-connected and stand-alone environments. Grid-connected EMS can also be known as system-level EMS. EMS comprises programs like demand response programs (DRP), demand-side management (DSM), power quality management, and energy efficiency. Three types of architecture based on control are decentralized, centralized, and hierarchical EMS.

- *Centralized EMS.* A central controller, along with a safe communication network and an efficient computing unit, forms the architecture of centralized EMS. An aggregator or utility may be a central controller. The information, such as consumption patterns of energy, meteorological data, and energy generation data of DERs, is collected from market operators about all the nodes. It is utilized to carry out optimization programs to achieve its goals and ensure smooth operation. This architecture provided optimal global performance. Centralized architecture also has some demerits; because all data are collected and managed in one location, there is an increased computational load, particularly when controlling many assets. This control structure becomes less useful for requirements involving real-time communication. Since each new source or component has its operating expenses and constraints, the centralized EMS is hard to expand and might get interrupted when integrating new sources or components. Also, it includes a single point of failure.

- *Decentralized EMS.* A distributed processing system where each node is capable of independent control and may communicate with other nodes peer-to-peer forms the architecture of decentralized EMS. The drawbacks of centralized architecture are overcome by decentralization, as it avoids single-point failure, enables expandability, and permits greater operational flexibility. Based on the level of decentralization and communication network availability, three modes of operation for decentralized EMS are fully dependent, fully independent, and partially dependent. In fully dependent mode, local controllers communicated through the central controller with each other. The central entity is involved for communication purposes, and it does not interfere in the decision-making process of the local controllers. In partial mode, the controllers interact with each other and also with the central entity. Communication among local controllers is independent of the central entity in a fully independent mode. Sufficient controllers and communication provide higher reliability in the decentralized architecture in comparison to the centralized one.
- *Hierarchical EMS.* This architecture is proposed for a microgrid community system. The system is subdivided into multiple levels of control. Information flows between adjacent levels and among units at the same level. The three control levels existing in this architecture are the optimization control level, supervisory control level, and execution control level.

Stakeholders and participants in EMS are system operators, ICT, advanced metering infrastructure (AMI), and aggregators. The literature in [39] clearly states that decentralizing batteries from centralized to distributed units resulted in better functionality. It suggests that decentralized batteries fully comply with reliability, scalability, and flexibility. When the operation is independent of central controllers, the failure of one individual unit does not affect the power system significantly. This reflects the improved reliability of the system. The number of units of batteries can be added or removed at any time according to the need; this increases the scalability. Similarly, flexibility is enhanced due to a variable number of batteries. As a result, decentralized residential batteries are being utilized by residential consumers.

Impact of ICT on Power Systems
Technological advancements have led to the development of ICT. Technologies that electronically transmit, process, store, generate, share, display, or exchange information are referred to as information and communication technologies by UNESCO [40]. They include both traditional technologies like radio and television and modern technologies like satellite communication, smartphones, computer hardware, and software. This is accompanied by applications and services and is enabling national and international bodies to decarbonize through digitalization.

The smart grid is also an application of ICT in power systems. Many of the problems the power industry faces are thought to have an answer in the smart grid. Essentially, a smart grid combines power systems and ICT, enabling bidirectional communication, and the combination of not only power but information is carried there, where a constant stream of intelligent devices communicate with one another.

The smart grid has enabled a transition from a centrally controlled environment to a more decentralized and consumer-interactive one [41]. It catalyzes the following benefits:

- Enabling the smooth integration of RESs like wind, and solar [17];
- Making ESS integration with the grid and residential consumers possible on grid-scale as well as at the residential level;
- Better utilization of RESs around the clock;
- Enabling ease of usage of plug-in EVs (PHEVs) as integration is eased; and
- Allowing consumers to participate effectively in DRPs, helping them better manage their consumption.

It also allows a transition from individual consumer grid interaction to consumer interaction in the form of the idea of grid-interactive efficient buildings (GEB). Figure 9.4 gives a schematic representation of GEB.

Currently, valuation, scheduling, implementation, and management of BTM-DERs are done independently, be it solar PV, EVs, ESS, etc. The integration and ongoing optimization of DERs for the grid's benefit as well as that of building owners and occupants is the goal of GEB [42].

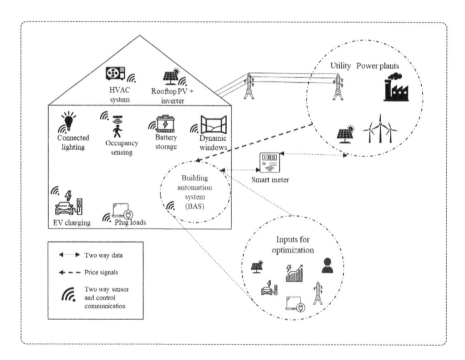

FIGURE 9.4   Grid-interactive efficient building.

9.4 ENERGY MANAGEMENT THROUGH RESIDENTIAL CONSUMERS' INVOLVEMENT

For residential consumers, batteries are a practically affordable storage solution, especially in distributed settings. When households first began integrating batteries, their primary goals were to use stored energy during periods of high electricity rates and to obtain power during power outages. The energy transition enabled them to become prosumers rather than just consumers. Hence, the purposes served by the BESS for decentralized residential consumers are not only limited to the benefits they can provide to battery owners, but they can potentially serve some of the system-level benefits. They serve purposes such as supply–demand balance, reactive power support, ancillary services in the form of frequency and voltage regulations, and peak shaving. Thus, they may in the future serve incredible benefits to the reliability, stability, resilience, and flexibility of the system, directed from the residential consumer side [25].

Now batteries are mostly being deployed along with PV rooftop solar. Li-ion batteries are most preferred by residential consumers lately, but this might change in the future based on technological advancements, changing market dynamics, and consumer preferences. The existing grid mix of generation, when paired with rising renewables, shows that overall system costs have been considerably lowered due to the flexibility offered by battery storage. Homeowners can participate in DRPs, where they supply stored energy to the grid during peak demand times, helping to alleviate stress on the electricity grid. This flattens the demand peak curve.

Energy storage systems have better performance potential when they are deployed closer to loads that are away from generation [43, 44]. For analyzing the benefits returned after deployment, stakeholders should consider the financial aspects of using decentralized residential batteries. The financial aspects to be considered while going for the adoption of decentralized residential batteries are as follows:

- *Initial investment.* This can include all the costs related to the purchase, installation, and required electrical upgrades. The determination of the actual and accurate price might depend on the size, type, and any extra components, say an inverter.
- *Operating and maintenance costs.* The costs related to regular checkups, repairs, warranties, and necessary maintenance are included in these costs.
- *Return on investment (ROI).* ROI determination is a critical financial consideration. This entails estimating the time it will take for future income (from demand response or grid services, for example) and savings on energy to offset the initial expenditure. In this case, it can depend on energy consumption patterns, electricity prices, and the availability of incentives or rebates. A positive ROI is expected out of the initial investment.
- *Incentives and rebates.* For motivating battery integration at the residential level, there is a need to take advantage of the opportunities provided. This reduces the burden of the initial investment and increases its financial feasibility.

- *Grid services and revenue generation.* If residential consumers have the opportunity to participate in DRPs or can export excess stored or self-generated energy back to the grid, this additional revenue can offset initial expenditures.
- *Energy policy and regulations.* Awareness of different national and local policies impacting the economics of decentralized residential batteries is important and can affect the decision on potential investments.

All the previous discussion has emphasized aspects related to combining decentralized batteries and residential consumers. The purposes that are served by BESS for residential consumers will be highlighted in the next section.

9.5   BENEFITS OF DECENTRALIZED RESIDENTIAL BATTERIES

ESS can be deployed at three levels: the transmission level, the distribution level, and BTM. The literature supports the idea that the ESS located in BTM is capable of offering more services to the system. It eventually facilitates system-level energy management. A schematic diagram showing the hierarchy of major stakeholders (groups) is shown in Figure 9.5. Major stakeholders (groups) that are affected by the deployment of BESS at the customers' side that is BTM are discussed in what follows [18, 44–47].

9.5.1   System-Level Benefits

The services offered to the system by the deployment of decentralized batteries behind the meter are as follows.

Customer Services

- *Time of use (TOU) bill management.* Behind the meter, consumers can deploy energy storage systems to minimize their costs; that is, they can reduce their electricity bill. They can do this by withdrawing power from the grid or purchasing power at lower TOU rates.
- *Increased PV self-consumption.* The value of self-consumption of electricity generated by BTM rooftop solar PV systems is typically much higher than the profit from exporting that electricity to the grid.

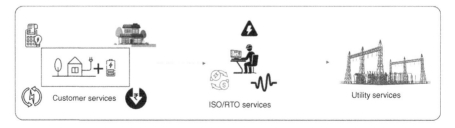

FIGURE 9.5   Hierarchy of benefits that are extended among major stakeholders (groups).

- *Demand charge reduction.* One can store (solar) energy that is produced so that it can be used at peak periods or whenever there is an energy spike by investing in a battery storage solution. That means pairing local generation with storage can greatly reduce demand charges.
- *Backup power.* Energy storage and a small generator can supply backup power at various scales in the case of grid breakdown, from second-by-second power quality maintenance for industrial activities to daily backup for residential consumers.

System Operation Services

- *Voltage support.* The regulation of voltage across the grid ensures a continuous and reliable electricity supply. To guarantee that both real and reactive power generation are matched with demand, the voltage on the transmission and distribution systems must be kept within an acceptable range.
- *Spin or non-spin reserves.* The spinning reserve stands for the generation capacity present online and is capable of serving loads immediately as a response to any unexpected or sudden contingency event, such as a sudden generation outage. The non-spinning reserve stands for generation capacity that is not instantaneously available but is capable of responding to contingency events in a typically short period (less than 10 minutes).
- *Energy arbitrage.* When wholesale electricity is purchased at the time of the lowest locational marginal price (LMP), typically during nighttime hours, it is sold back to the wholesale electricity market at the time of the highest LMP.
- *Black start.* Black start generation assets are required to restart larger power plants in the case of a grid outage to put the regional power grid back online. Large power plants are capable of carrying out black starts in certain situations.
- *Frequency regulation.* When any change in system frequency is locally sensed (in the system or any of its elements), the automatic and immediate response of power is taken care of by frequency regulation.

Utility Services

- *Distribution deferral.* When investments by utilities in the direction of distribution system upgrades or updates are postponed, scaled back, or completely avoided as they were needed to meet the anticipated load increase in a particular region of the grid, this is known as distribution deferral.
- *Transmission deferral.* When investments by utilities in the direction of transmission system upgrade or updating are postponed, scaled back, or completely avoided as they were needed to meet the anticipated load increase in a particular region of the grid, this is known as transmission deferral.
- *Transmission congestion relief.* For the use of busy transmission corridors at specific periods of the day, ISOs charge utilities. To discharge during busy times and reduce congestion in the transmission system, assets such as energy storage can be placed downstream of crowded transmission corridors.

- *Resource adequacy.* Consumption of electricity during peak hours puts forward the need to invest in new generation assets or technologies to meet that demand. Grid operators and utilities can rather invest in other assets, such as energy storage, which can reduce the requirement for new generation capacity. This reduces the risk of overinvestment.

9.5.2 COST–BENEFIT ANALYSIS

In order to determine whether moving towards decentralized residential batteries would bring more value and profit to the stakeholders involved in this process, a cost–benefit analysis will be useful. A general cost–benefit analysis of decentralized residential batteries can depend on many factors, including the location of deployment, regulatory and policy conditions, and individual circumstances. The overall profit to customers is based upon several elements, including energy consumption, regional electricity prices, and the availability of the programs that provide grid services. Decentralized residential battery storage can result in savings along with enhanced energy security. Depending on how much decentralized household batteries contribute to demand response, grid stability, and cost savings relative to the costs enabling integration into the grid and incentives, utilities may or may not get a net profit. Nevertheless, the potential benefits in terms of DSM and grid stability frequently exceed the disadvantages. The case studies of Australia and India, which are discussed in the later section, support this analysis.

9.6 CASE STUDIES

A significant change is being seen in terms of energy management strategies across the globe. The distribution network is witnessing the entry of new elements like BTM energy storage systems. This is because large investments have been made in the direction of transferring some generation capacity to the end user's site, mainly for people using rooftop solar PV in the distribution network. The benefits offered by this setup give economic signals to both end users and grid utilities [35].

Key drivers for end users can be their increased interest in power quality, reliability, energy supply security, and self-consumption, as well as their ability to keep up with the changing dynamics of pricing mechanisms. BTM-ESSs are installed in many countries. This includes the Netherlands, the United Kingdom, Japan, the United States, and Australia.

This chapter will offer one case study from Australia and another from India. The results of these studies imply that governments can support or continue to support schemes that push a nation towards a greener future. These rollouts should be accompanied by the implementation of necessary policy directives.

9.6.1 AUSTRALIA

Australia has utilized its solar potential to a great extent by investing a lot in BTM PV (rooftop) along with energy storage systems to drift towards net zero targets [23].

Hence, it is included as a case study on how the installation of residential batteries led to behavioral changes in consumers.

AMI has played an important role in providing data for the analysis. Data from more than 5000 consumers with different arrangements of DG, like PV and ESSs, were taken into account. The data used were energy consumption by consumers and exported energy recorded by the smart meter.

Economic benefits for the installation of ESS (BESS) are put forward in terms of the IRR and PBP for the owners. Additionally, these storage systems also provided a wide range of benefits to the distribution network [45].

This case study provides two main pieces of evidence. First, it recommends the distribution network operator (DNO) based on the data, and then it details the cost–benefit analysis of residential battery installations. The three major categories of consumers were as follows:

- No PV-No Battery (nP-nB): 1428 consumers
- PV-No Battery (P-nB): 1567 consumers; the size of the PV system is not known
- PV-Battery (P-B): 2297 consumers
  - 1120 consumers; PV system size is known with an average of 5.7 kW
  - 140 consumers; battery size known with an average of 4.5 kW

House demography (family members, ages, financial situations, etc.) and properties of batteries (SOC or life span, etc.) were not known to the network providers. Hence, the data do not provide any of the above information. Network providers have no control over the batteries, which means they are decentralized residential batteries. In order to determine whether storage systems are affordable for consumers, one must understand the economics involved in installation and operation. Now, parameters considered indicative of the affordability of BESS were the PBP and IRR. The study provided us with the following conclusions:

- The PV-Battery (P-B) system was more beneficial to consumers as the energy consumption was reduced by 46 percent on average. This reduction reduced peak demand.
- The PBP for batteries lasting more than 25 years indicates the technical immaturity of the technology, making it unaffordable for residential consumers. These consumers, if incentivized continuously in the form of subsidies or discounts, should keep going until the technology becomes mature and affordable.
- The PV-battery combination is capable of offering encouraging opportunities for residential self-sufficiency if the right sizes of PV and battery, along with sufficient incentives, are provided.
- DRPs could extract greater benefits when batteries are used by residential consumers.

9.6.2 INDIA

A techno-economic analysis of a residential PV-battery system was conducted to find out profitability for a system located in Kannur, Kerala, India. The 5 kW PV-battery, along with a grid-export facility, was considered for the study in the mentioned region [48]. Li-ion batteries were taken into consideration for this analysis, and a net metering system was used. The parameters thus calculated were net present value, PBP, discounted PBP, and profitability index. The calculations indicated that PV, along with the battery, was profitable for the owner. Not all the locations with this combination can bring in profits. Better subsidies and schemes can increase profits for both consumers.

9.7   CHALLENGES

Challenges that can persist even after the deployment of decentralized residential batteries are as follows.

- Batteries will get charged during excess generation, but after a battery is fully charged, the excess generation during the daytime will spill back to the grid, causing a voltage disturbance.
- Differential tariffs can result in generating peaks other than the real peak. High electricity prices during peak hours shift the peak towards times of cheaper electricity prices.
- The high cost and shorter life span of residential batteries hinder them from becoming an economically viable solution.
- Current market conditions might not support investments in this direction without financial support.
- Benefits derived after these installations might be able to compensate for the investment cost within the lifespan of the installed battery, which is the lengthy PBP.
- The Li-ion battery market for static storage applications might face competition with the EV market. There is no prediction available regarding the impacts of this competition on static storage.

The above-stated challenges often pose problems for distribution network operators. This demotivates consumers from installing rooftop solar PV, as the excess generation during the day cannot be exported to the grid. Thus, this cannot encourage the deployment of large-scale and complex businesses. Lack of data can even lead to an immature business case, which will demotivate investments in this direction. Thus, there is a need to work on these issues, considering suitable business models.

9.8   GOVERNMENT INITIATIVES IN INDIA TO SUPPORT BESS

In September 2023, the Indian government rolled out a scheme to support the development and construction of BESS assets for delivery by 2030–2031 [49]. The government foresees the award of funding to support around 4000 MWh of BESS resources. This aims to reduce the cost of battery energy storage for both consumers and distribution companies. The government has approved viability gap funding (VGF) with up to 40 percent of the capital cost of the projects. This scheme provides VGF funding and aims to achieve an LCOS of between Rs. 5.50 and Rs. 6.60 per kWh, which will make stored renewable energy a practical solution for controlling peak electricity demand throughout the nation. Distribution companies (Discoms) shall be granted access to a minimum of 85 percent of the BESS project capacity to guarantee that the scheme's advantages are realized for the customers. A fair and competitive bidding process will be used to choose BESS developers for VGF funds, ensuring equal opportunities for both private and public sector participants.

The scheme aims to give consumers access to green, clean, reliable, and reasonably priced electricity by utilizing the potential of RESs. The very mandate of competition is fulfilled by the competitive bidding process, as it will promote healthy competition as well as lead to the development of a strong ecosystem for BESS. As a result, the need for infrastructural development will be deferred. This will not only support renewable integration but will also lead to optimum utilization of the transmission network.

9.9 CONCLUSION

In this chapter, an effort has been made to investigate the potential benefits of using decentralized residential batteries to achieve system-level energy management. The adoption of battery energy storage systems at the distribution side, closer to loads, has potential benefits. The major stakeholder groups that are most affected by the deployment of behind-the-meter storage are consumers, utilities, and system operations. The hierarchy of benefits starts at the consumer's end. Consumers get continuous power, privacy is protected, and they save money through bill management, increased PV self-consumption and generation, exporting to the grid, peer-to-peer trading, and demand charge reduction. They can participate in DRPs to take advantage of incentives or rebates provided by agencies. Thus, willingness to participate in the market and contribute to system-level benefits increases. This leads to benefits for system operators. The penetration of DER causes grid imbalances. Load peaks result in voltage fluctuations, phase imbalances, and ohmic losses. Thus, peak shaving helps in mitigating these challenges. System operators face fewer grid imbalances. They also earn profits because of reduced voltage and frequency imbalances, increased energy arbitrage, and fewer black starts. They save on ancillary services. Batteries installed nearer to loads end up in transmission and distribution network deferral. Self-generation and consumption lead to fewer imports from the grid, reducing congestion, and helping utilities earn profits. Resource adequacy is gained because of the deployment of battery energy storage systems. The risk of overinvestment gets reduced as an investment in new infrastructure is deferred. Moreover, case studies from Australia and India made it evident enough that the initial cost of investment can be high while adopting renewable energy along with batteries, but the value of the savings on the electricity bill is much higher. Households that installed PV plus batteries earned the most profit out of all the three scenarios considered. This setup even reduced peak demand. Depending on market dynamics, consumption patterns, and technological advancements, the battery energy storage system's needs might change.

The main hindrances to the adoption of decentralized residential batteries are the high initial cost of deployment and short lifetime. Incentives and rebates can ease the adoption of batteries, along with rooftop solar PV, in households or communities.

REFERENCES

[1] G. Cubuk, "Comparison of emissions gap reports by building sector between 2014–2022 through data analysis," *Journal of Environmental Protection and Ecology,* vol. 24, no. 2, pp. 397–407, 2023.

[2] N. Yu, W. Wang, and R. Johnson, "Behind-the-meter resources: Data-driven modeling, monitoring, and control," *IEEE Electrification Magazine*, vol. 10, no. 4, pp. 20–28, 2022.

[3] I. A. Fernandez, D. Ramasubramanian, W. Sun, A. Gaikwad, J. C. Boemer, S. Kerr, and D. Haughton, "Impact analysis of DERs on bulk power system stability through the parameterization of aggregated DER_a model for real feeders," *Electric Power Systems Research,* vol. 189, p. 106822, ISSN 0378-7796, 2020.

[4] R. R. Trivedi, C. P. Barala, P. Mathuria, R. Bhakar, and S. Sharma, "Peer-to-Peer energy trading: Energy pricing using game theory models," in *Conference of 2023 IEEE IAS Global Conference on Renewable Energy and Hydrogen Technologies (GlobConHT)*, Male, Maldives, pp. 1–6, 2023.

[5] J. C. Peña, C. C. Zarate, A. P. Duque, and G. O. Pinto, "Distributed energy resources on distribution networks: A systematic review of modelling, simulation, metrics, and impacts," *International Journal of Electrical Power & Energy Systems*, vol. 1, no. 138, p. 107900, 2022.

[6] X. Liu, X. Liu, Y. Jiang, T. Zhang, and B. Hao, "Photovoltaics and energy storage integrated flexible direct current distribution systems of buildings: Definition, technology review, and application," *CSEE Journal of Power and Energy Systems*, vol. 9, no. 3, 2023, pp. 829–845.

[7] S. R. Sinsel, R. L. Riemke, and V. H. Hoffmann, "Challenges and solution technologies for the integration of variable renewable energy sources—A review," *Renewable Energy*, vol. 145, pp. 2271–2285, 2020.

[8] S. Kakran and S. Chanana, "Smart operations of smart grids integrated with distributed generation: A review," *Renewable and Sustainable Energy Reviews*, vol. 81, pp. 524–535, 2018.

[9] R. Hull and A. Jones, "Development of decentralised energy and storage systems in the UK," A Report for the Renewable Energy Association, KPMG, 2016.

[10] N. Shaukat *et al.*, "Decentralized, democratized, and decarbonized future electric power distribution grids: A survey on the paradigm shift from the conventional power system to microgrid structures," *IEEE Access*, vol. 11, pp. 60957–60987, 2023.

[11] A. J. Ernst, I. Ansah, K. Bachus, C. Kporxah, and O. S. Tomomewo, "The role of energy storage in the evolution of renewable energy and its effect on the environment," *American Journal of Energy Research*, vol. 11, no. 3, pp. 128–143, 2023.

[12] Statista Research Department, "Electricity consumption in the United Kingdom (UK) from 1970 to 2022, by final user (in terawatt-hours)," UK Government, BEIS, 2023.

[13] N. Horesh, "Economic and environmental evaluation of emerging electric vehicle technologies," Colorado State University, 2023.

[14] Eurostat. "Energy use in households up 6% in 2021 [online]," 2023. Available: https://ec.europa.eu/eurostat/web/products-eurostat-news/-/ddn-20230613-1

[15] Government of India, National Statistical Office, "Chapter 6: Consumption of energy resources," in *Energy Statistics India 2023,* GOI, MoSPI, pp. 49–64, 2023.

[16] S. Agnew and P. Dargusch, "Consumer preferences for household-level battery energy storage," *Renewable and Sustainable Energy Reviews,* vol. 75, pp. 609–617, 2017.

[17] D. Huber, D. Costa, A. Felice, P. Valkering, T. Coosemans, and M. Messagie, "Decentralized energy in flexible energy system: Life cycle environmental impacts in Belgium," *Science of the Total Environment*, vol. 886, p. 163882, 2023.

[18] C. Jankowiak, A. Zacharopoulos, C. Brandoni, P. Keatley, P. MacArtain, and N. Hewitt, "Assessing the benefits of decentralised residential batteries for load peak shaving," *Journal of Energy Storage*, vol. 32, pp. 101779, 2020.

[19] J. Mitali, S. Dhinakaran, and A. A. Mohamad, "Energy storage systems: A review," *Energy Storage and Saving*, vol. 1, no. 3, pp. 166–216, ISSN 2772-6835, 2022.

[20] M. Krichen, Y. Basheer, S. M. Qaisar, and A. Waqar, "A survey on energy storage: Techniques and challenges," *Energies*, vol. 16, pp. 2271, 2023.

[21] H. Ibrahim, A. Ilinca, and J. Perron, "Energy storage systems—Characteristics and comparisons", *Renewable and Sustainable Energy Reviews*, vol. 12, no. 5, 1221–1250, 2008.

[22] O. Palizban and K. Kauhaniemi, "Energy storage systems in modern grids—Matrix of technologies and applications," *Journal of Energy Storage*, vol. 6, pp. 248–259, 2016.

[23] M. Rezaeimozafar, R. F. Monaghan, E. Barrett, and M. Duffy, "A review of behind-the-meter energy storage systems in smart grids," *Renewable and Sustainable Energy Reviews*, vol. 164, pp. 112573, 2022.

[24] International Hydropower Association, "2022 Hydropower status report sector trends and insights," Report, IHA, 2022.

[25] M. S. Mastoi, S. Zhuang, H. M. Munir, M. Haris, M. Hassan, M. Alqarni, and B. Alamri, "A study of charging-dispatch strategies and vehicle-to-grid technologies for electric vehicles in distribution networks," *Energy Reports*, vol. 9, pp. 1777–1806, 2023.

[26] B. Bibak and H. Tekiner-Moğulkoç, "A comprehensive analysis of Vehicle to Grid (V2G) systems and scholarly literature on the application of such systems," *Renewable Energy Focus*, vol. 36, pp. 1–20, ISSN 1755-0084, 2021.

[27] M. Kwon and S. Choi, "An electrolytic capacitorless bidirectional EV charger for V2G and V2H applications," *IEEE Transactions on Power Electronics*, vol. 32, no. 9, pp. 6792–6799, 2016.

[28] B. Zakeri, G. C. Gissey, P. E. Dodds, and D. Subkhankulova, "Centralized vs. distributed energy storage–benefits for residential users," *Energy*, vol. 236, pp. 121443, 2021.

[29] T. Bowen, I. Chernyakhovskiy, and P. Denholm, "Grid-scale battery storage," NREL, Golden, CO (USA). NREL/TP-6A20-74426, 2019.

[30] M. Noussan, "Economics of electricity battery storage," in *The Palgrave Handbook of International Energy Economics*, M. Hafner and G. Luciani, Eds. Palgrave Macmillan, Cham, 2022, pp. 235–253.

[31] I. S. Bayram and T. S. Ustun, "A survey on behind the meter energy management systems in smart grid," *Renewable and Sustainable Energy Reviews*, vol. 72, pp. 1208–1232, 2017.

[32] J. P. Carvallo, N. Zhang, S. P. Murphy, B. D. Leibowicz, and P. H. Larsen, "The economic value of a centralized approach to distributed resource investment and operation," *Applied Energy*, vol. 269, pp. 115071, 2020.

[33] International Renewable Energy Agency, "Solar energy capacity, installed solar energy capacity," IRENA, 2022.

]34] International Renewable Energy Agency, "Renewable capacity statistics," IRENA, Abu Dhabi, 2023.

[35] R. Boampong and D. P. Brown, "On the benefits of behind-the-meter rooftop solar and energy storage: The importance of retail rate design," *Energy Economics,* vol. 86, p. 104682, 2020.

[36] P. Graham and L. Havas, "Projections for small-scale embedded technologies," Report, CSIRO, Australia, 2020.

[37] M. Schulze, H. Nehler, M. Ottosson, and P. Thollander, "Energy management in industry – a systematic review of previous findings and an integrative conceptual framework," *Journal of Cleaner Production,* vol. 112, no. part 5, pp. 3692–3708, ISSN 0959-6526, 2016.

[38] S. K. Rathor and D. Saxena, "Energy management system for smart grid: an overview and key issues," *International Journal of Energy Research,* vol. 44, no. 6, pp. 4067–4109, 2020.

[39] A. Reindl, H. Meier, and M. Niemetz, "Scalable, decentralized battery management system based on self-organizing nodes," in Conference *International Conference on Architecture of Computing Systems*, Cham, Springer International Publishing, pp. 171–184, 2020.

[40] C. Patel, M. Dubey, and K. Sinha, "Linkages of information and communication technology (ICT) and the Indian diaspora," *Transdisciplinary International Journal of Academic Research*, vol. 1, pp. 6–14, 2023.

[41] K. Chavan and R. Guhagharkar, "Role of ICT in power system", Study committee D2 information systems and telecommunication, 2013.

[42] M. Neukomm, V. Nubbe, and R. Fares, "Grid-interactive efficient buildings", Technical report for the US Department of Energy, 2019.

[43] A. M. Nour, A. A. Helal, M. M. El-Saadawi, and A. Y. Hatata, "Voltage imbalance mitigation in an active distribution network using decentralized current control," *Protection and Control of Modern Power Systems*, vol. 8, no. 1, pp. 1–17, 2023.

[44] G. Fitzgerald, J. Mandel, J. Morris, and H. Touati, "The economics of battery energy storage: how multi-use, customer-sited batteries deliver the most services and value to customers and the grid," Rocky Mountain Institute, 2015.

[45] N. Al Khafaf, A. A. Rezaei, A. M. Amani, M. Jalili, B. McGrath, L. Meegahapola, and A. Vahidnia, "Impact of battery storage on residential energy consumption: An Australian case study based on smart meter data," *Renewable Energy*, vol. 182, pp. 390–400, 2022.

[46] C. Jankowiak, A. Zacharopoulos, C. Brandoni, P. Keatley, P. MacArtain, and N. Hewitt, "The role of domestic integrated battery energy storage systems for electricity network performance enhancement," *Energies*, vol. 12, no. 20, p. 3954, 2019.

[47] A. O. Abbas, B. H. Chowdhury, "Using customer-side resources for market-based transmission and distribution level grid services – A review," *International Journal of Electrical Power & Energy Systems,* vol. 125, p. 106480, ISSN 0142-0615, 2021.

[48] G. Krishnan and K. V. Chandrakala, "Techno-economic analysis for the profitability of residential PV-battery system," in Conference *2022 4th ICEPE*, Shillong, IEEE, April 2022, pp. 1–5.

[49] Government of India, Ministry of Power, "National framework for promoting energy storage systems," New Delhi, MOP, 2023.

10 Current Challenges and Application Outlook of Battery Technologies in Home Energy Systems

Phuc Duy Le and Duong Minh Bui

10.1 INTRODUCTION

The rapid advancement of renewable energy sources and the increasing need for energy storage solutions have made battery technology for home energy systems become a hot topic. This chapter investigates recent problems with battery technology and how it might be applied in households. A number of battery technologies utilized in home energy systems will be introduced in the first section. Section 10.2 then discusses the challenges that battery technology in home energy systems must face. These difficulties include high costs, limited energy density and capacity, and environmental problems with battery manufacturing and end-of-life management. The chapter delves into the need for more robust safety approaches, enhanced operational efficiency, and ecologically sustainable manufacturing methods to surmount these challenges. The possible applications of battery technology in home energy systems are next examined in Section 10.3, which highlights the potential of batteries in enabling higher self-consumption of renewable energy, enhancing grid resilience through energy storage devices, and aiding load shifting and peak shaving. The chapter explores the potential benefits and considerations for homes of integrating battery technology with solar panels, smart home systems, and infrastructure for charging electric vehicles. The chapter also lists present advancements and new trends in battery technology, such as solid-state, flow, lithium-ion, and others. It deliberates about how home energy systems could be able to overcome obstacles and take advantage of new opportunities with the aid of new battery technologies. In order to overcome the challenges and completely utilize battery technology in household energy systems, the chapter's conclusion emphasizes the significance of further research, innovation, and collaboration among stakeholders, policymakers, and researchers. By removing these constraints, battery technologies have the power to completely change the way homes use and manage energy, heralding a more reliable and sustainable energy supply in the future. In summary, this chapter sheds insight on the challenges that battery technologies recently face in home energy systems while

DOI: 10.1201/9781003441236-10

also highlighting the promising future of these technologies. It makes clear how battery innovations contributed to the transition to more efficient, environmentally friendly home energy systems.

10.2 BATTERY TECHNOLOGIES IN HOME ENERGY SYSTEMS

Batteries are an essential component of household energy storage and utilization systems. Extra energy generated by renewable resources, such as solar panels, wind turbines, and off-peak utility grid hours, may be able to be stored thanks to them. This chapter expresses that the following important factors need to be taken into account (see Table 10.2).

- Firstly, lithium-ion batteries are widely used in home energy systems due to their high energy density, extended cycle life, and efficiency. Their durable design and lightweight make them suitable for usage in domestic settings. Over time, lithium-ion batteries have grown more affordable as a reliable choice for energy storage at home.
- Secondly, it is crucial to consider both capacity and power requirements when choosing a battery system for a residence. Whereas capacity refers to the total amount of energy the battery can retain, power controls how quickly the battery can deliver energy. Depending on their needs, users can give priority to either more capacity or better power output.
- Thirdly, depth of discharge (DOD) is the proportion of a battery's total capacity that is used up before recharging. For example, an 80 percent DOD battery can be discharged to 80 percent maximum of its capacity. Deeper discharges can reduce the longevity of a battery system; hence it is important to consider its DOD.
- Fourthly, the percentage of a battery's total capacity that indicates how much energy is still in the battery is called the state of charge (SOC). It shows the current state of the battery, including how full or empty it is. The state of charge of the battery indicates its energy level or capacity relative to its maximum capacity. The most frequent way to represent it is as a percentage, where 100 percent indicates that the battery is fully charged and 0 percent indicates that it is completely exhausted.
- Fifthly, energy management systems are often included in home energy systems to optimize battery consumption. These systems monitor energy generation, consumption trends, and battery charge levels to guarantee efficient use of energy. It can self-charge the battery when renewable energy output is at its peak or when demand is low.
- Finally, battery safety needs to be the first priority for in-home energy solutions. Lithium-ion batteries frequently have built-in safety procedures including thermal management systems and battery management systems (BMS) to prevent overheating, overcharging, and overdischarging. Regular maintenance, such as monitoring performance and ensuring enough ventilation, is necessary to maximize battery lifespan and efficacy.

It is crucial to remember that the accessibility and particular features of battery technologies can vary depending on the location and the products provided by regional producers and suppliers. It is advised that consumers considering installing a battery-operated home energy system consult with local professionals or energy service providers who are able to assess their particular needs and provide reasonable recommendations. It is also significant to realize that energy storage systems, or ESS, have multiple positive effects on the electric grid. Increasing energy efficiency, reducing greenhouse gas emissions, and improving grid stability are a few possible benefits. It is essential to assess ESS's multiple applications and related benefits from the perspectives of various stakeholders in order to maximize these benefits. Examining the wholesale and retail energy market regimes might potentially yield profitable outcomes for the numerous ESS applications. Selecting the appropriate storage technology for a given application is also essential. Presenting and comparing storage technology selection procedure might result in a definite technological choice even when there are capable candidate storage systems.

Application-specific control strategies are needed for an ESS to function under both balanced and unbalanced grid environments. The ESS can review and describe various control mechanisms in order to operate effectively and efficiently. Last but not least, it is critical to remember that integrated applications managed by an ESS may provide better operational and financial results than stand-alone applications. Understanding and recommending this usage will help to maximize the benefits of ESS in the power system. To sum up, this chapter provides valuable insights into the applications and benefits of energy storage systems (ESS), which may be utilized to understand and investigate the overall benefits of ESS for the electrical grid.

10.2.1 ELECTROCHEMICAL ESS

Chemical energy is transformed into electrical energy by electrochemical power sources. Electric current is available at a specific voltage and time as the reaction's energy. Electrochemical batteries and electrochemical capacitors are the two main branches. A battery is a device that uses an electrochemical oxidation-reduction (redox) reaction, which involves the movement of electrons from one material to another via an electric circuit, to directly convert the chemical energy included in its active materials into electric energy. The actual electrochemical device that produces or stores electric energy is called a cell. Batteries are made up of one or more cells linked in series, parallel, or both, depending on required output voltage and capacity. A comparison of certain main and secondary battery cell details can be found in Table 10.1.

An investigation of the many battery kinds reveals that each type has a unique set of uses. On the other hand, primary cells have been found to offer more benefits in terms of large loads and long-term preservation. In addition to being more durable than other battery kinds, primary cells are also less expensive to produce and maintain the system as a whole. They thus offer a more dependable substitute for energy storage devices that have an ongoing need to operate. When integrating an ESS into the bigger electrical grid, consideration must be given to the costs and long-term

TABLE 10.1
Comparison of a battery's primary and secondary cells

Primary cells	Secondary cells
Use electrolytes instead of rechargeable materials (absorbent substance or separator)	Rechargeable by flowing a current into the circuit that is opposite to the discharge current
Two categories: aqueous and non-aqueous	There are two kinds: non-aqueous (solvents) and aqueous (water).
Lower starting price	A higher starting price
A higher cost-per-kWh life cycle	A reduced life cycle cost ($/kWh) if charging is affordable and convenient
Disposable	
A replacement is easily accessible	Consistent maintenance and recurring recharging are necessary
Usually more compact and lighter, making them better suitable for portable applications.	Although they are available, replacement batteries are not manufactured in the same volume as primary batteries; preordering could be required
Higher charge retention and longer service per charge	Conventionally less suitable for portable applications, although new developments in lithium battery technology have made secondary batteries lighter and smaller
Not the best option for high performance/ heavy load/high-rate discharge	
Imperfect for high-priced military applications, load-leveling, emergency backup, and hybrid batteries	Traditional secondary batteries, especially aqueous secondary batteries, have lower charge retention when compared to primary battery systems
Historically restricted to particular uses	Exceptional performance at high discharge rates under heavy loads
	Perfect for hybrid batteries, load-leveling, emergency backup, and expensive military applications
	The inherent versatility of secondary battery systems as a whole facilitates study and use for a wide range of applications

benefits of the various types of batteries. Primary cells are especially suitable for this use since they store energy more efficiently and need less preservation than other types of batteries. They are therefore the greatest option for energy storage systems that have long durations of full capacity operation. All things considered, employing primary cells in energy storage systems can have a number of benefits that outweigh those of other battery types. The kind of battery can be accurately chosen to create a more efficient, dependable, and cost-effective energy storage system that can support the broader electric grid (Table 10.2).

TABLE 10.2
Comparison of different battery types

Type	The traditional battery system: Zinc alkaline manganese dioxide	The vehicle battery: Lead-acid battery system	The laptop battery: Lithium battery system	The torpedo battery: Silver oxide battery system
Application	Radio, calculator, remote control, regular camera, compact disc player, cellular phone, etc.	Automobile engine starting, lighting and ignition, energy storage, emergency power, electric/hybrid vehicles, submarines, etc.	Laptops, advanced cellular phones, personal data assistants, military electronics (mine detectors, satellites, military radios, thermal weapon sights)	Torpedoes, aquatic mines, swimmer aids, deep submersibles, underwater rescue vessels, various antisubmarine warfare applications
Cathode anode electrolyte	Manganese dioxide/zinc/ aqueous potassium hydroxide	Lead dioxide/metallic lead/sulfuric acid solution	Lithium cobalt oxide/graphite/ lithium hexafluorophosphate (most common)	Silver oxide/zinc/aqueous potassium hydroxide
Distinguishing characteristics	Greater density of energy Improved customer service Reduced internal resistance Extended shelf life Higher leakage resistance Increased stability in dimensions	Common, inexpensive secondary battery, globally produced Accessible in an array of sizes Strong, peaky performance Electrically efficient, with a 70% turnaround efficiency The highest voltage (>2.0V) at which an aqueous electrolyte battery system can be operated Excellent float service East charge state indication For applications requiring intermittent charging, good charge retention Available in styles that require no maintenance Cheaper than comparable solutions for secondary batteries Cell parts can be recycled with ease	Sealed batteries that require no maintenance Extended life of cycle Extended shelf life Capable of functioning in a wide variety of temperatures Low rate of self-discharge Quick recharge ability Capacity for high-rate and power discharge Excellent energy and coulombic efficiency High energy density and specific energy No retention impact	Elevated energy in relation to weight and volume Capacity for both high and moderate discharge rates Strong retention of charges A flat voltage curve for discharge Low maintenance Decreased self-discharge Secure

(continued)

TABLE 10.2 (Continued)
Comparison of different battery types

Type	The traditional battery system: Zinc alkaline manganese dioxide	The vehicle battery: Lead-acid battery system	The laptop battery: Lithium battery system	The torpedo battery: Silver oxide battery system
Additional information	This kind of battery has supplanted the Leclanche battery system as the standard battery for consumer devices in recent years because of its better performance at lower temperatures and larger current discharges, along with a longer shelf life.	Annual sales of lead-acid batteries represent approximately 45% of the sale volume of all batteries in the world.	Lithium has the lightest weight, highest voltage, and greatest energy density of all metals.	Zinc/silver oxide batteries provide the highest energy per unit weight and volume of any commercially available aqueous secondary battery system. A significant portion of the early research on the zinc/silver oxide battery technology was sponsored by the US Navy.

10.2.2 CAPACITOR AND ELECTROCHEMICAL CAPACITOR

Classical Capacitor

- The most straightforward way to store energy.
- Comprised of a dielectric layer, which acts as a non-conducting barrier between two metal plates.
- A direct current source causes one plate to become charged, while electromagnetic induction causes the other plate to become charged in the opposite direction.
- Metal electrodes or metalized plastic sheet have surfaced that store energy.
- Low energy density allows for the rapid delivery or acceptance of high currents.

Super-Capacitor

- Also referred to as an ultra-capacitor, electric-double layer capacitor, or electrochemical capacitor (EC).
- Employs an electrolyte layer that is as thin as a molecule.
- Possesses an activated carbon structure with an extremely big surface area.
- Superior energy storage capacity to traditional capacitors.
- Able to produce a high output of peak power.
- Has a lengthy life cycle and can be recharged and discharged millions of times without damage.

10.2.3 COMPARISON BETWEEN BATTERY AND CAPACITOR

Given that they both store and release electrical energy, batteries and capacitors appear to be comparable. But because of the ways in which they operate differently, there are important distinctions between them that affect their possible uses (Table 10.3).

Batteries and capacitors do have certain overlapping uses, despite their significant distinctions. But generally speaking, capacitors can charge and discharge more rapidly than batteries, and batteries can store more energy density (greater power density). Researchers are working to lengthen the battery's charging and discharging durations and expand the capacitor's storage capacity in response to the market's need for quick portable power. Batteries and capacitors still have unique qualities that make them suitable for specific applications even as research into improving them continues. The most viable option for ESS is batteries. Electrochemical devices, such as batteries, may easily transform stored energy into electrical energy. They do not need to take into account geography because they are movable and frequently relatively small. Only rechargeable batteries are of interest for large-scale energy storage; batteries can be classified as primary or secondary depending on whether they are non-rechargeable or rechargeable. Additionally, batteries might be flow or solid-state batteries.

TABLE 10.3
Summary of comparative results between battery and capacitor

	Batteries	Capacitors
Pros	Density of power	Extended life of cycle
	The capacity to store	High load currents
	Leakage current superior to capacitors	Quick charging periods
	A constant voltage with on/off switches	Outstanding temperature performance
Cons	Short cycle life	Low energy-specificity
	Limitations on current and voltage	Linear voltage of discharge
	Prolonged charging periods	Excessive discharge
	More temperature sensitive than capacitors	Price per watt is expensive

10.3   RELATED CHALLENGES

Even though household energy systems (HES) battery technology has advanced significantly, certain issues still need to be resolved. The following are some significant constraints that battery technology in household energy systems must overcome.

- *Investment cost (installation, operations, and maintenance).* For many households, the prepaid cost of BESS might be a major obstacle. Although the price of lithium-ion batteries has come down over time, it is still rather expensive when considering alternative energy storage solutions. Some individuals might be put off by the upfront cost of a battery-powered home energy system. In this chapter, this feature will be meticulously studied.
- *Limited lifespan.* The number of cycles (charge and discharge) that batteries can withstand before needing to be replaced is limited. Over time, a battery's capacity deteriorates, reducing its total energy storage capacity. A battery's life can be affected by DOD, temperature, and charging and discharging rates. Replacing batteries can be more expensive for homeowners.
- *Safety concerns.* Battery safety is tremendously significant when working with larger energy storage devices. Even though lithium-ion batteries have safety characteristics, improper maintenance increases the risk of thermal runaway, overheating, or fire. It is essential to abide by safety rules, use appropriate installation and maintenance techniques, among other things, in order to reduce these risks.
- *Environmental impact.* Batteries can have an effect on the environment through their manufacture, application, and removal. The environment may be harmed during the extraction and processing of raw materials for batteries. Moreover, to improve the impact of wasted batteries on the environment, effective recycling and disposal are needed.
- *Integration with the utility power grid.* The integration of HESs with the electrical grid may face certain technical obstacles. Effective integration of BESS with the grid requires that they be able to communicate with it and respond to signals from it, such as load shifting or demand response programs. Verifying

the compatibility and efficiency of the home energy system and the grid might be difficult.

- *Regulatory and policy frameworks.* HESs and battery technologies may be different depending on the regional laws and policies. Policies concerning grid access, net metering, tariffs, subsidies, and other related issues may have an impact on the adoption and economics of home energy systems. The integration of battery technologies can be facilitated and greater usage can be encouraged by streamlining regulations and offering supportive policies.

It is important to note that attempts are being made to overcome these issues as battery technologies continue to progress. The goals of ongoing research and advancement are to enhance battery time of use, cost-effectiveness, performance, and safety. Government incentives and supportive policies can also aid in removing obstacles and encouraging the broad use of battery-powered household energy systems.

10.3.1 Depth of Discharge

The percentage of a battery's total capacity that has been used before recharging is known as the depth of discharge (DOD). It shows how much of a battery has been used up in comparison to its total capacity. A battery with a 50 percent DOD, for example, has been depleted to half of its total capacity. DOD directly affects how long to use a battery. Deeper discharges can shorten a battery's cycle life by hastening the deterioration process. Capacity fading may occur more noticeably in batteries that are depleted to a larger proportion of their overall capacity over time.

The process of charging and discharging a battery is referred to as battery cycling. The lifespan of the battery can be increased by using a shallow cycling approach, which involves discharging the battery to a lower DOD before recharging. Preventing deep discharges lessens the battery's pressure, which delays capacity loss and lengthens its cycle life. The DOD has different suggestions for different battery chemistries in order to enhance their lifespan. Lithium-ion batteries, for instance, typically function between 20 and 80 percent DOD to balance energy storage requirements while reducing the effect on battery lifespan.

It is crucial to take the intended DOD into consideration when designing a HES with energy-based batteries needs and battery parameters. If extending the battery's longevity is the top concern, a shallower cycle technique with a lower DOD should be used. To efficiently meet energy demands, nevertheless, a greater battery capacity might be necessary. Battery management systems (BMS) are essential for overseeing and managing the DOD. They avoid over-discharge, which can harm the battery, and make ensuring it runs within safe parameters. The BMS protects the battery from damaging substances and enhances battery performance. It is important to keep in mind that specific DOD recommendations could vary depending on the battery chemistry, manufacturer specifications, and intended use. When selecting the optimal DOD for the user battery system, refer to the battery manufacturer's instructions or seek advice from prestigious installers or specialists. The DOD of a battery is simply expressed as a percentage in Equation (10.1):

$$DOD = (Discharged\ Capacity\ /\ Total\ Capacity) * 100\%\qquad(10.1)$$

where *Discharged Capacity* is the amount of energy discharged from the battery, and *Total Capacity* is the battery's maximum energy storage capacity. Divide the discharged capacity by the total capacity, then multiply the quotient by 100 to find the DOD. For instance, the DOD can be computed as follows if a battery with a 1000 amp-hour (Ah) total capacity has discharged 750 amp-hours:

$$DOD = (750\ Ah\ /\ 1000\ Ah) * 100\% = 75\%$$

This means that 75 percent of the battery's total capacity has been used up.

10.3.2 STATE OF CHARGE

An important consideration when mentioning battery energy systems is the state of charge (SOC). In terms of a battery's total capacity, it expresses the remaining power as a percentage. At any given time, it provides information about how full or empty the battery is. When a battery is nearing its maximum capacity, its state of charge is indicated by its SOC. Typically, it is expressed as a percentage, with 100 percent denoting a fully charged battery and 0 percent denoting a totally discharged battery.

Monitoring a battery's SOC is crucial for efficient use and battery safety. By preventing deep discharges and overcharging and ensuring that the battery is used within safe operating ranges, it helps users make the most efficient use of energy. Battery systems often incorporate indicators or user interfaces that display the SOC in order to provide consumers with real-time information. These interfaces could be in the form of mobile apps, digital displays, or other visual signals. They allow users to monitor the battery's energy level and make informed decisions about how to use and charge it.

Many methods can be used to detect and monitor a battery's SOC. Battery monitoring systems (BMSs) use voltage, current, and temperature measurements to accurately calculate the SOC. The devices' real-time information on the battery's energy level allows users to make informed decisions about how best to use it. The SOC also varies with the battery's state of charge or discharge. The SOC increases when energy is stored in the battery and decreases when energy is withdrawn.

The accurate SOC of a battery may be hard to verify due to numerous factors, such as temperature effects, internal resistance, and battery lifespan. While BMSs provide estimates, they may not always be precise. The accuracy of SOC estimation can be improved by calibration and regular examination with pre-established reference points. Batteries often have recommended operating parameters so that the SOC can guarantee optimal performance and endurance. These parameters depend on the battery chemistry and manufacturer's specifications. As stated by the manufacturer, the battery's life can be extended and its efficiency can be improved. The SOC and DOD are two different concepts. The SOC displays the remaining energy as a percentage of the battery's total capacity, whereas the DOD indicates how much of the battery has been used or depleted. It is essential to comprehend this differential while controlling and monitoring the battery's energy levels. The SOC needs to be tracked and controlled for efficient battery utilization and protection. It ensures that the battery is operated within safe operating parameters, allows users to use energy more effectively, and avoids deep discharges and overcharges. With the increasing use of BESS, it is more important than ever to monitor and control the SOC of batteries. It enables users to choose wisely, improve performance, and extend battery life.

To sum up, in order to maximize the lifespan and performance of batteries, it is imperative to monitor and control their state of charge. Reliable SOC predictions are provided by battery monitoring systems, and the accuracy of SOC estimations can be increased by calibration and recurring examination with established reference points. Users can make sure that the battery operates at its best, extend its longevity, and utilize it within safe operating limits by paying attention to the SOC and adhering to the manufacturer's instructions. It is more important than ever to monitor and manage the SOC of batteries due to the growing use of battery energy systems. To calculate the SOC percentage, Equation (10.2) can be used:

$$SOC(t) = \frac{Q_{remaining}(t)}{Q_{max}(t)} \times 100 \, [\%] \qquad (10.2)$$

While it might not be simple to obtain for every chemistry, the SOC is essential for all applications and provides the foundation for the calculation of other states. The relationship between the SOC and the voltage measured in lithium-ion batteries—and LFP in particular—is extremely non-linear, necessitating the employment of prediction and estimation methods to determine it.

In general, the number of cycles determines the battery capacity. The average SOC and the SOC swing determine how many operating cycles there are. The battery's voltage, current, efficiency, and charging/discharging time all have a significant impact on the SOC level. However, the battery voltage is the most crucial factor in determining the SOC level and the BM failure rate under dynamic operating circumstances.

10.3.3 END OF LIFE

When a battery reaches its end-of-life (EOL) point, it is no longer functional or able to achieve the required performance standards. The battery's appropriateness for the intended application may be diminished at this point if it is unable to deliver the desired capacity or power output. Important factors to take into account when it comes to battery EOL are listed below.

First, one of the main causes of battery EOL is capacity fade. Frequent cycles of charging and discharging cause the battery's capacity to steadily decline over time. As a result, the battery's capacity to store and release energy has decreased from its original specifications. Capacity fading is influenced by a number of parameters, such as battery chemistry, usage habits, operating conditions, and depth of discharge. The anticipated lifespan of a battery technology is determined by the number of cycles of charging and discharging that the battery can withstand before it reaches the end of its useful life. For instance, a lithium-ion battery's lifespan could reach several thousand cycles, contingent upon its chemistry and operational circumstances. The expected battery lifespan is the most important factor to take into account when designing and implementing a home energy system.

A battery's performance might decline in various ways as it gets closer to the end of its life. Higher internal resistance in addition to declining capacity might cause the battery to produce less power and operate less efficiently. The system's overall performance may be impacted by this degradation, necessitating battery replacement or refurbishment. But for safety and environmental reasons, treating EOL batteries carefully is essential. Lithium-ion batteries are among the several battery technologies that have potentially dangerous components that need to be handled carefully. Numerous locations provide battery recycling programs in an effort to salvage valuable materials while reducing the negative effects of battery disposal on the environment. It depends on the laws and guidelines in local area regarding the proper recycling or disposal of batteries.

There are situations where batteries with reached the end of their useful life for a particular application can still be helpful. Secondary or "second-life" applications can be found for these batteries. An electric car battery that is no longer appropriate for use in automobiles, for instance, can nevertheless have sufficient capacity to be utilized for stationary energy storage. Before batteries are recycled or disposed of, second-life uses might increase their overall useful life. The development of more recyclable and sustainable battery chemistry, as well as improvements to durability and lifespan, are all part of the ongoing efforts to advance battery technology. Furthermore, advancements in battery recycling technology are being made to enhance the effectiveness and environmental impact of managing end-of-life batteries. As a result, the details of EOL management can change based on the chemistry of the batteries, regional laws, and the infrastructure for recycling that is in place. Lastly, in order to guarantee that batteries are handled and disposed of properly in accordance with local regulations, it is advised to consult local specialists or recycling facilities.

A variety of characteristics and conditions are routinely monitored and analyzed in order to determine the EOL of BESS. Although there are no set formulas for calculating EOL, a number of indicators and computations can be used to predict the remaining useful life and degradation of batteries. In accordance with their equations and metrics, a number of widely used approaches are available, such as coulomb counting, Peukert's law, cycle count, state of health, and voltage and impedance monitoring:

$$SOH = \frac{Q_{max}(t)}{Q_{nominal}} \times 100[\%] \tag{10.3}$$

where SOH stands for the battery's current capacity in relation to its initial capacity. A percentage is frequently used to express it.

If we use the cycle count approach, a cycle is usually defined as one full charge and discharge cycle of a battery. Subsequently, tracking and documenting the amount of cycles can aid in estimating battery life. The relationship between a battery's capacity, runtime, and discharge rate is expressed as follows in Peukert's law:

$$T = \frac{C}{I_{discharging} \times \left(\dfrac{C}{Q}\right)^k} \tag{10.4}$$

where k is the Peukert exponent, Q is the Peukert capacity constant, I is the discharge current, T is the runtime, and C is the capacity.

As stated in Equation (10.4), the coulomb counting approach integrates the current over time to estimate the capacity used or remaining in a battery. Accurate current measurement and knowledge of the battery's initial state of charge—which is often updated upon full charge—are essential components of this estimating technique. Even when this is executed correctly, the CC method keeps integrating faults brought on by imprecise measurements and other factors, which may eventually lead to a significant loss of method reliability.

$$SOC(t) = SoC_0 - \frac{1}{C_{nom}} \times \int_{t_0}^{t} i_{batt}(t) \, dt \times 100 \tag{10.5}$$

where I is current, and t is time.

Variations in voltage and impedance characteristics in the voltage and impedance monitoring method can reveal information about battery deterioration and a gradual drop in performance. This straightforward approach correlates the battery's internal voltage with its state of charge and uses a lookup table with predetermined values. Because it is so basic, it is irrelevant to online applications because the only way to determine the internal voltage of a battery is to disconnect it and wait for it to stabilize.

The ratio of peak power to nominal power is known as the battery's state of power, or SOP. The maximum power that can be sustained constant for T seconds without going over predetermined operational design limitations on battery voltage, SOC, power, or current is known as the peak power, depending on the current state of the battery pack.

$$SOP(t) = \frac{P_{max}(t)}{P_{nominal}} \times 100 \, [\%] \tag{10.6}$$

This indicator is crucial for ensuring that the battery is used to its full potential and that its life expectancy is extended by ensuring that the charge or discharge power does not exceed set restrictions. Additionally, this signal might be helpful in peak power applications in determining battery conditions that allow for large charges or discharges. The SOP is acquired in a second step of a battery study because it is extremely dependent on the SOC, the battery's capacity, chemistry, and voltage.

10.3.4 CAPITAL INVESTMENT

The amount of capital needed for a home-integrated battery energy system can vary based on a number of variables, such as the brand, installation difficulty, power demand, battery capacity, and any extra equipment necessary. The following are the primary costs to bear in mind:

- *Battery system cost.* One of the biggest expenses of the capital investment is the battery itself. The decline in battery prices over time can be attributed mostly to advancements in lithium-ion battery technology and increased manufacturing outcomes. Nevertheless, depending on the capacity and brand, the price might still be high, ranging from a few thousand to tens of thousands of dollars.
- *Inverter/converter.* Inverters or converters are frequently needed in battery systems in order to change the direct current (DC) that is stored in the battery to the alternating current (AC) that is used by household appliances. A user might have to buy an additional inverter or an inverter with integrated battery management depending on the current setup. Inverters range in cost depending on their features and capacity.
- *Installation costs.* When installing a battery system, professional services like electrical work, system integration, and even authorization costs are sometimes needed. The complexity of the installation and the particular requirements of the user's home may have an impact on the installation price. It is critical to consult with trustworthy installers who can offer accurate cost estimates based on the layout of the user's home.
- *Balance of system components.* Additional components can be needed, depending on the user's design, to connect the battery system to their current electrical infrastructure. Wire, circuit breakers, monitoring systems, and any upgrades or modifications to the user's electrical panel are a few examples of these components. User capital investment should take these system components' costs into account.
- *Maintenance and operational costs.* The costs associated with battery care and operation must be taken into account, even though they are not included in the initial capital investment. It is feasible to do software updates, periodic inspections, battery capacity tests, and, if required, component replacement over time. The performance and lifespan of the battery system can be increased with proper maintenance.
- *Incentives and rebates.* Government incentives, tax credits, or rebate programs may be available to help offset the initial cost of a battery system, depending on the user's location. To find possible cost-saving measures, it is advisable to study local programs or consult experts because these incentives vary by area and are subject to change.

It is important to note that long-term advantages like decreased dependency on the grid reliance, possible savings on energy costs, and more energy independence might result from the initial investment made in a battery system. Additionally, carefully consider the user's energy requirements, financial situation, and potential return on investment before making a choice. It is advisable to work with prestigious suppliers or energy system installers who can provide precise quotes based on a thorough assessment of the user's household requirements in order to achieve a more accurate estimate of the capital expenditure needed for a battery system adapted to the user's specific needs. The general formula for a capital investment of a home-integrated BESS is displayed in Equation (10.7).

$$
\begin{aligned}
Capital\ Investment = \ &Battery\ System\ Cost + Inverter\ and\ Power\ Electronics\ Cost + \\
&Installation\ and\ Labor\ Costs + Balance\ of\ System\ Costs \\
&+ Site\text{-}Specific\ Costs - Government\ Incentives \\
&+ Maintenance\ and\ Warranty\ Costs \qquad (10.7)
\end{aligned}
$$

10.4 APPLICATION OUTLOOKS WHEN INTEGRATING BESS AT HOME

10.4.1 SOURCES AND LOAD OPTIMIZATION

Integrating BESS at home can offer several application outlooks that provide benefits to homeowners. Some potential applications and advantages of integrating BESS at home are listed as the following:

1. One of a BESS's main uses is to supply backup power in the event of distribution grid failures. A precisely constructed and configured battery system can plug in automatically to maintain vital operations like communication, lighting, and refrigeration as well as power main appliances in the event that the main grid fails. This lessens interruption-related discomfort and improves energy resiliency.

2. Another way BESS can help manage energy usage is through time of use optimization, which involves storing electricity during off-peak hours when prices are usually lower. The total cost of energy can be decreased by using the stored energy during peak hours, when electricity rates are higher. For this application, the battery operation needs to be intelligently and automatically controlled by the EMS in order to maximize savings. Demand charges are applied to electricity bills in certain areas, and they are determined by the maximum power used during periods of peak demand. Homeowners can lower their peak power draw from the utility grid and hence their demand charges on their payments by deliberately draining the battery during times of peak demand. BESS enables homeowners to change their energy consumption behavior to more sustainable and clean sources by storing excess RES. This leads to a greener, more ecologically friendly energy usage profile and less reliance on fossil fuel-based grid electricity.

3. Integrating a BESS allows solar energy to be consumed on its own by homes with installed solar panels. When solar generation is low, excess energy from the sun can be stored in batteries for use at a later time or to lower relying on

the utility grid. This can maximize the utilization of self-generated renewable energy and increases energy independence.

4.   In periods of high demand or unpredictable supply, BESS can support the grid by taking part in demand response and grid services. To reduce the stress on the main grid, load shifting—in which the BESS is charged during off-peak hours and discharged during peak demand hours—can be used. Homeowners who take part in these programs may be able to obtain incentives or payments.

It is important to remember that the benefits and application outlooks of integrating battery storage systems can change based on a number of factors, including the battery capacity, household energy requirements, local electricity pricing, and available incentives or programs that are offered in the user's area. To ascertain the best uses and possible advantages for user-specific home energy systems, consult with respectable installers, energy specialists, or system integrators.

BESS as a Backup Power Source
When demand for electricity exceeds supply, a BESS is a device that stores electrical energy in rechargeable batteries and provides power during blackouts or when electricity demand exceeds supply. By giving homeowners a stable backup power supply, this lessens their dependency on the main power grid and guarantees that essential appliances and systems will always have power. The capacity of a battery, commonly measured in kilowatt-hours (kWh), indicates how much electrical energy it can store. This quantity represents the total energy that can be used up while the battery is operating on backup power.

In order to better benefit from this service, the BESS is equipped with an inverter that converts DC electricity stored in the batteries into AC electricity, which powers household appliances and systems. The inverter, which also controls electricity flow, ensures a smooth transition between grid power and backup power. In order to provide real-time monitoring of the battery's SOC, system performance, and other critical factors, the BESS also includes a monitoring and control system. It allows homeowners to keep updated on the state of the battery, make economical use of energy, and manage backup power operations. To ensure a smooth and continuous transition to backup energy, an automated transfer switch is a device that detects main grid power failures and automatically switches the power source from grid power to the backup power system.

Depending on their energy needs, homeowners may be able to add additional battery capacity to BESS through expandability. Because of its flexibility, homeowners can customize their backup power system to meet their particular requirements. For homeowners, having a BESS as a backup power source has several benefits. Utilizing stored energy during periods of high demand or high electricity prices encourages energy independence, strengthens resilience during blackouts, and can lower electricity costs. Figure 10.1 shows a BESS for homes that includes PV and loads. As demonstrated in this video, the house receives power not just from the utility grid but also from its batteries and solar panels that are powered by a centralized inverter. The owner can keep an eye on this inverter using a local computer or a website that uses a cloud-based platform.

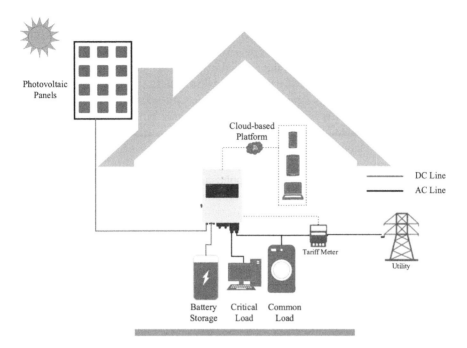

FIGURE 10.1 A household-based BESS with PV and loads.

Enabling TOU Optimization Capabilities

Time of use (TOU) optimization capabilities in a BESS refer to its capacity to judiciously regulate the battery's charging and discharging based on fluctuating electricity tariffs throughout the day. This optimization aims to save electricity costs during off-peak hours by minimizing reliance on the more costly peak hours of the grid. See the items below for an explanation of how the TOU optimization works.

When electricity prices are lower during off-peak hours, the BESS can be configured to charge overnight. By storing energy during off-peak hours and using it during peak hours, when rates are higher, homeowners can reduce the overall cost of their electricity consumption. Furthermore, the TOU optimization feature allows the battery system to transition from using grid power to stored energy during peak-rate periods. An EMS needs to be integrated with the BESS in order to track electricity costs in real time and determine when to charge or discharge the battery. The EMS takes into account factors like the battery's SOC, past consumption patterns, and current electricity pricing in order to optimize the utilization of stored energy over different time periods. Moreover, the smart grid system can incorporate a complex BESS with TOU optimization capabilities. As a result, the system can adjust its charging and discharging patterns as needed and obtain up-to-date pricing information for electricity. It enables the battery to respond to grid signals, optimize energy use in response to pricing signals, and take grid conditions into account.

By using TOU optimization skills, homeowners can enjoy lower electricity rates, reduced grid demand during peak hours, and even lower energy costs. The system intelligently schedules the charging and discharging cycles of the battery to align with the most cost-effective periods and typical energy consumption patterns. It is crucial to remember that the utility and accessibility of the TOU optimization capabilities may vary depending on the particular BESS and the corresponding energy management software. Advise from an expert or a battery storage provider to learn more about the TOU optimization capabilities of specific systems and how they might be customized to meet specific needs. Reduced reliance on grid power, lower electricity rates, and potential energy cost savings are just a few advantages of numerous advantages when having TOU optimization capabilities in a BESS. Homeowners can maximize energy use throughout the day by intelligently controlling the battery's charging and discharging cycles. The application of BESS in reducing peak load during the day and shifting it to the night is shown in Figure 10.2.

Surplus PV Power Absorption
The use of extra solar power produced by a PV system that is not instantly consumed by a home's or building's electrical load is known as surplus PV power absorption. Surplus PV power can be fed back into the electrical grid in systems that are connected to grid. These are referred to as feed-in tariff or net metering programs. The homeowner receives credit or payment for the exported energy when the PV system produces more electricity than the house uses. The excess power is returned to the grid. This allows the effective use of excess PV power and encourages the grid's integration of renewable energy sources.

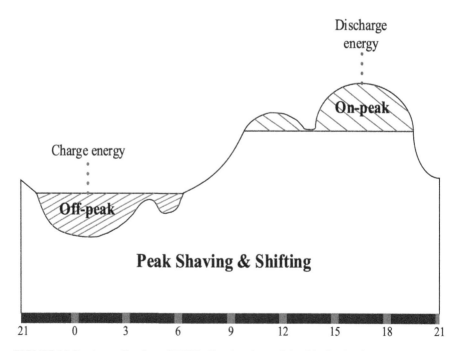

FIGURE 10.2 An application of BESS: Shaving the peak load in the daytime and shifting to nighttime.

Local restrictions prevent homeowners in certain areas from sending back to the grid. As a result, to maximize the effectiveness and advantages of solar energy, excess power from a photovoltaic system that is not being used can be stored or absorbed. Storing extra photovoltaic electricity in batteries for later use is one such method. A battery energy storage system, or BESS, has the ability to store surplus electricity produced during the day and release it when solar generation is minimal or non-existent, like in the evening or on cloudy days. As a result, households can depend less on the grid and more on solar energy that has been stored. Furthermore, extra electrical loads or appliances in the house can be powered by excess PV electricity. The excess power can be directly absorbed by energy-intensive equipment, such as electric water heaters, electric vehicle chargers, or pool pumps, by strategically timing the operation of these devices to coincide with periods of high solar generation, minimizing dependency on grid electricity. Figure 10.3 illustrates how more power is generated from renewable sources than is used at home. The BESS can use that extra generation power during the day to absorb it and use it to reduce the peak demand at night. In the event of capacity limitations, Figure 10.4 illustrates two methods for homeowners to avoid the penalty fine.

Easier to Access the Demand Response Program
Demand response (DR) programs have used BESS in a variety of contexts. By providing incentives for consumers to cut back on their electricity use or move it to off-peak hours, demand response (DR) programs seek to control and balance the amount of electricity consumed during peak hours. By supplying stored energy during times of high demand or absorbing extra electricity during times of low demand, BESS can be a useful tool in demand response. In DR programs, three common applications, namely load shifting, peak shaving, and ancillary services, are used efficiently.

1. *Load shifting.* BESS can be utilized to move energy usage to off-peak times. When there is less demand for electricity or when electricity prices are lower, the BESS can be charged with excess renewable energy production or grid power. In order to minimize the need for grid-supplied electricity, the battery's

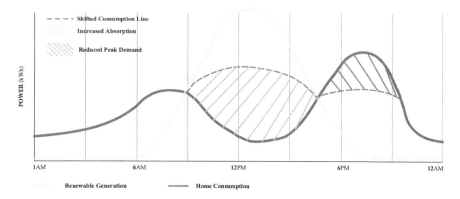

FIGURE 10.3   A demonstration of an effective approach to utilize surplus PV generation.

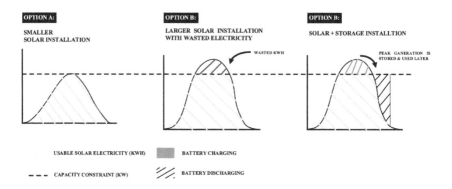

FIGURE 10.4   Two options for utilizing surplus PV generation by applying BESS at home.

stored energy is able to be used during periods of high demand to power the house or facility.

2. *Peak load shaving.* BESS can be used to reduce the peak electricity demand by supplying power during periods of high demand. The battery supplements and discharges the electrical load instead of using the grid, lessening the grid and avoiding peak-time charges. This lowers the need for additional electricity output during peak hours and helps to stabilize the system.

3. *Ancillary services.* Grid stability, reactive power support, and voltage control are just a few of the ancillary services that BESS can provide to the grid. These services can be especially helpful when there is grid stress or a system outage since they maintain the quality and stability of the energy supply.

Grid operators and utilities can profit from the adaptability and responsiveness of stored energy by incorporating BESS into DR operations. Better control of electricity demand, load balancing, and grid stability is made possible by the use of BESS for DR. Additionally, users that participate in demand response programs could be given financial rewards or discounted electricity costs, as shown in Figure 10.5.

10.4.2   Electric Vehicles and Electric Vehicle Charging Stations

Electric vehicles (EVs) are revolutionizing the automotive sector, revolutionizing transportation, and paving the way for a more sustainable future. The rise of EVs in the global market can be attributed primarily to advancements in technology, increased environmental awareness, and the transition to renewable energy sources. The introduction of this chapter highlights their key characteristics, benefits, and impacts on society and the environment. Rechargeable batteries power one or more motors in EVs, as opposed to internal combustion engines (ICEs), which rely on fossil fuels for operation. For EVs, electricity is the primary energy source. It can come from a number of sources, including renewable energy sources like hydropower, wind, and

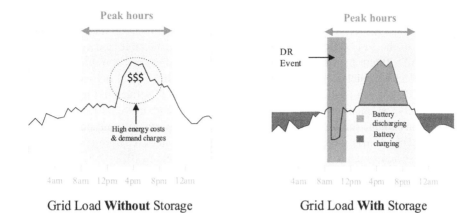

Grid Load **Without** Storage   Grid Load **With** Storage

FIGURE 10.5 A BESS discharges its storage power in a DR event to avoid the high energy costs and demand charges.

sun. Among the main advantages of EVs are their environmental benefits. By getting rid of the exhaust emissions from traditional internal combustion engines, EVs significantly lower air pollution and greenhouse gas emissions, contributing to the fight against climate change and improving the quality of air in cities. Furthermore, by lowering reliance on finite fossil fuel resources, the use of EVs promotes energy diversification and decreases dependency on imported oil.

An essential component of EVs is their energy efficiency. Electric motors are more efficient than internal combustion engines, converting a larger percentage of electrical energy into real vehicle propulsion. Because of their efficiency, owners of EVs consume less energy, pay less for operating costs, and possibly even witness a decrease in the demand for energy overall. Infrastructure for batteries and charging EVs has been improving alongside EV technology. Modern EVs use lithium-ion batteries, which have greater energy densities, longer driving ranges, and quicker charging times. A full network of charging infrastructure, comprising fast-charging stations next to highways, public charging stations, and home charging stations, has also contributed to the widespread adoption of EVs. EVs offer benefits for the environment and energy consumption in addition to a unique driving experience. Because of their smooth acceleration and instant torque, EVs provide a quiet and comfortable ride. The consumer appeal of EVs is enhanced by the incorporation of cutting-edge features and technologies like autonomous driving capabilities, smart connection, and regenerative braking.

However, it is critical to acknowledge the challenges and worries surrounding EVs. Concerns about battery production and recycling, infrastructure accessibility and availability for charging, and the price and range limitations of specific EV models are a few of these. Governments, industry stakeholders, and consumers must keep up their research, innovate, execute laws that support it, and collaborate to address these problems. EVs present a practical solution for reducing air pollution, mitigating climate change, and enhancing energy efficiency within the transportation sector. With the development of new infrastructure, laws that favor them, and ongoing technological advancements, EVs are poised to have a significant influence on transportation in the future. Societies can shift to a more sustainable and environmentally friendly future by switching to electric vehicles. There are now four different types of electric vehicles: ICE, hybrid electric vehicles (HEV), plug-in hybrid electric vehicles (PHEV), and battery electric vehicles (BEV), as illustrated in Figure 10.6.

EVs rely on many essential components to run and move. Together, these components enable the use of electric propulsion, energy storage, and other vehicle features. The electric motor, battery pack, power electronics, charging system, and auxiliary systems are the key components, as shown in Figure 10.7.

1. The main means of propulsion in an EV is an electric traction motor. It transforms the battery's electrical energy into mechanical energy, which powers the wheels. The motor can have several configurations, such as induction motors or permanent magnet motors, and can be either AC or DC.

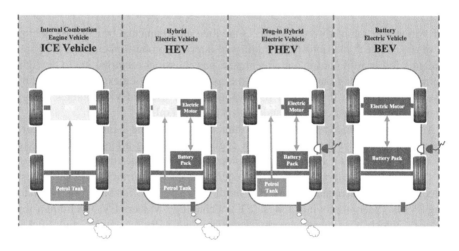

FIGURE 10.6   State-of-the-art electric vehicle categories.

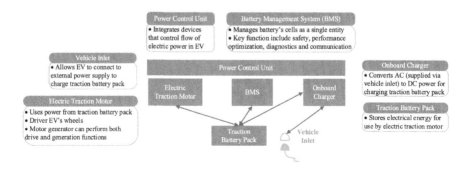

FIGURE 10.7   Key EV elements.

2. The traction battery pack is the part of an EV that stores energy. It is composed of many individual batteries with the capacity to store electrical energy. The most common battery type used in EVs is lithium-ion, which has a high energy density, decent efficiency, and a competitive long lifespan. The vehicle's electric motor and other systems are powered by the battery pack.

3. The amount of electricity that is transported between the battery pack and the electric motor is managed by power conversion devices. They accomplish this by transforming the battery pack's DC power into the AC energy needed by the electric motor and by controlling the voltage and current flow to guarantee safe and efficient operation.

4. The EV's traction battery pack can be recharged using the onboard charger by using an external power source. Usually, it consists of an onboard charger that converts wall socket or charging station AC power into DC current suitable for charging the battery pack. Different charging levels, such as level 1 (120V AC), level 2 (240V AC), and level 3 (high-power DC fast-charging), are supported by EVs.

5. Battery management systems (BMSs) support EV operation and enhance the driving experience. Examples include the power steering system, air conditioning and heating systems, braking system, lights, infotainment systems, and safety features. They also can manage the power injected into the vehicle's battery pack or other dedicated power sources.

6. The vehicle inlet allows EV to connect to the external power source from the house's socket or EV charging station to charge the traction battery pack.

Enabling dependable and effective charging procedures requires effective communication between EVs and EV charging stations (EVCS). Standardized protocols and interfaces enable information exchange and charging session coordination between the car and the charging station. The ratings and charging time information for EV communications are displayed in Table 10.4. Also, there are a few essential elements of EV-EVCS communication that need to be understood.

1. *Plug and connector standards.* The physical connection between the EV and the EVCS is made possible by standardized plug and connector standards. Various plug and connector standards, including Type 1 (SAE J1772) and Type 2 (IEC 62196) connectors, have been adopted by various regions and countries to guarantee compatibility and safe electrical connections between the car and the charging infrastructure.

2. *Communication protocols.* Information sharing between the EV and the charging station is made possible by communication protocols. The open charge point protocol (OCPP), an open-source standard designed to make communication between EVCSs and back-end management systems, is one of the protocols that is frequently used. Features like authorization, management over charging sessions, and authentication are enabled by OCPP.

3. *Authentication and authorization.* EVs and EVCSs authenticate and authorize charging sessions via communication protocols. This guarantees that the infrastructure for charging can only be accessed and used by authorized users.

TABLE 10.4
EV communications regarding ratings and charging time

Plug	Communication	Current	Power	Charging time	Voltage
Type 1 SAE J1772	Pulse width module	≤16 Ampere	1.9 kW	Up to 8 hours	Single-phase 120V$_{AC}$
		≤80 Ampere	19.2 kW	Up to 8 hours	Single-phase 240V$_{AC}$
CHAdeMO	CAN	≤400 Ampere	400 kW	No more than 1 hour	1000 V$_{DC}$
Type 2	Pulse width module	≤32 Ampere	7.4 kW	Up to 8 hours	Single-phase 230V$_{AC}$
		≤63 Ampere	43 kW	Up to 8 hours	Three-phase 400V$_{AC}$
CCS/Combo	Basic: pulse width module High level: PLC	≤500 Ampere	500 kW	No more than 1 hour	1000 V$_{DC}$

Mobile apps, RFID cards, and other secure-identifying systems are examples of authentication methods. By confirming user credentials, the charging station has the authority to grant or refuse access to the charging service.

4. *Charging parameters and status.* EVs and the locations of wall connection charging stations exchange data about the charging session's characteristics and general state. Consequently, this includes the battery's current state of charge, charging power level, charging rate, and predicted charging duration. These data are transmitted to the EVs by the charging station, enabling the car to modify its charging behavior according to the power available and the user-specified charging preferences.

5. *Billing and payment.* Payment and billing procedures are also communicated between EVs and EVCSs. The length of the charging session, the amount of energy used, and the related cost can all be found out using EVCSs. The EV receives this information, which may be utilized for billing purposes. The charging station can be equipped with integrated payment methods, allowing customers to pay for charging services with credit cards, smartphone apps, or other safe payment methods.

6. *Smart grid integration.* The infrastructure of the smart grid can be integrated with EVs and EVCSs thanks to advanced communication capabilities. Features like load control, demand response, and bi-directional power flow (also known as vehicle-to-grid, or V2G) become possible through this. EVs can contribute to grid-balancing programs, return electricity to the system during times of peak demand, and help maximize the use of renewable energy resources by interacting with the grid.

In order to ensure compatibility and interoperability between various EV models and providers of charging infrastructure, standardized protocols and interfaces are essential. This helps to encourage the wider use of electric vehicles and the development of charging infrastructure. A seamless charging experience, improved charging efficiency, and assistance for grid integration and administration are all ensured by effective communication between EVs and EVCSs.

Temperature, battery SOC, battery capacity, and charging power are the four variables that determine how long an EV takes to charge. First, as Equation (10.8) illustrates, the charging power is determined by the rated power that an EV can obtain and that an EV charging station can offer.

$$Charging\ power\ in\ kW = min\{on-board\ charger\ rating, charging\ station\ rating\}$$
(10.8)

Battery capacity is the second component, meaning that a larger capacity will store more energy and require a longer time to fully charge. Furthermore, if the batteries' SOC is run at a temperature between 20 and 80 percent, they might charge more quickly. Lastly, the battery may be greatly impacted by the temperature. For example, in cold weather, batteries charge more slowly because the electrical flow is slowed down by the cold. Equation (10.9) can be used to calculate the predicted charging duration:

$$Estimated\ charging\ duration = \frac{Energy\ run\ into\ batteries}{Charging\ power}$$
(10.9)

An electric vehicle supply equipment (EVSE), sometimes referred to as an EVCS or electric vehicle charging station, is a specialized infrastructure created to provide electric power for EV battery charging. These stations offer the electrical connection and charging capacity required for recharging the battery packs of electric vehicles, including buses, motorbikes, and cars. To meet varied charging needs, EVCSs are available in a variety of types and charge levels. Popular EVCS types include the following:

- The most basic form of EV charging, known as EVCS level 1, entails utilizing the supplied charging cable to connect the car to a regular 120-volt AC household outlet. Although level 1 charging is usually slower than higher charging levels, it can be useful for charging overnight or in situations when a faster charge is not necessary.
- Compared to level 1, EVCS level 2 stations operate at a greater voltage (often 240 volts AC) and allow faster charging rates. For these stations to work, a specific charging device must be installed in a house, office, business area, or public space. An electric vehicle's battery can be fully charged in a few hours using level 2 charging, which is frequently utilized for daily usage.
- EVCS level 3 stations are DC fast-charging stations that offer powerful charging capabilities to quickly charge the batteries of electric vehicles. Bypassing the inbuilt charger of the car, these stations provide DC power to the vehicle.

Typically, users can find these DC EVCSs near public charging stations, major roadways, and highways. Depending on the particular charging rate and battery capacity, they can charge an EV to a substantial portion of its capacity in between 30 and 60 minutes.

Standardized connectors and communication protocols are frequently found on EVCSs to guarantee interoperability with various EV models. Additional characteristics that charging stations may have include energy management capabilities, billing and payment systems, authentication methods, and interaction with smart grid infrastructure. In order to encourage the adoption and use of electric vehicles, it is essential that charging stations be accessible and available. This will help the expanding electric vehicle market by offering a reliable and convenient charging infrastructure. The internal key components of AC and DC EVCS, as well as the EVCS principles, are depicted in Figures 10.8–10.10, respectively.

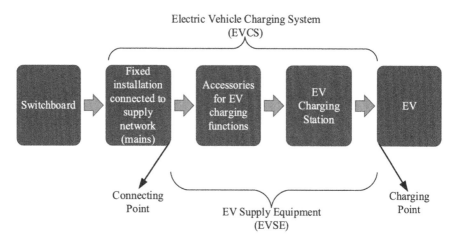

FIGURE 10.8   The EVCS concept.

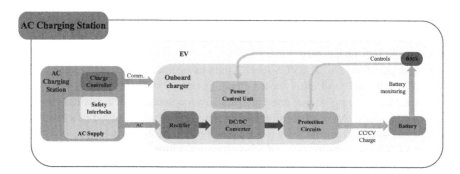

FIGURE 10.9   Key EV elements of the AC EVCS.

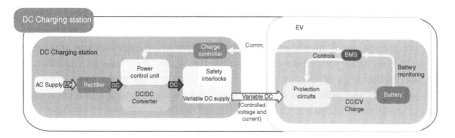

FIGURE 10.10 Key EV elements of the DC EVCS.

At-home BESS and EVCS integration can provide a number of benefits and extra features. Users can take use of a number of cutting-edge features that can optimize energy consumption, save energy expenses, and improve energy resilience by merging these two systems. Energy optimization is one of the main benefits of combining a BESS and an EVCS. Your home's energy usage, including the electricity required to charge your EV, can be optimized with a BESS. The excess energy can be retained in the battery rather than going straight to the EVCS during times when there is a decrease in the demand for power or when renewable energy generation (like solar energy) surpasses immediate needs. The battery can release its stored energy later on when user EVs need to be charged, which will lessen the load on the grid and possibly utilize self-generated renewable energy.

Load management is another benefit of combining a BESS and an EVCS. An additional load may be placed on the electrical grid when charging an EV at home, particularly if several EVs are charging at once. Because the battery can supply surplus power during EV charging, lowering the peak power pull from the grid, integrating a BESS enables load management. This can be helpful if you have limited electrical capacity or want to avoid excessive demand charges because it balances the whole home's electricity consumption and eliminates dangerous overloads. A BESS can assist in maximizing the usage of grid electricity for charging consumers' EVCS in areas where time of use electricity pricing is in place.

Combining a BESS and an EVCS can offer backup power for users' homes and EVCSs during grid failures, in addition to energy optimization, load control, and grid interaction. Your EV and other vital home appliances, as well as crucial loads like them, can continue to receive electricity from the battery during a power outage, guaranteeing your power and mobility in an emergency. Lastly, centralized control and management are made possible by the integration of a BESS and an EVCS. The BESS and EVCS can be monitored and controlled by users through specialized software or a smart EMS. This makes it possible to effectively schedule, optimize, and coordinate charging and discharging operations to meet user preferences, grid circumstances, and energy demands. To make sure the BESS and EVCS are appropriately matched, it is crucial to take their capacities into account. Furthermore, to guarantee correct installation, electrical compatibility, and best system performance when integrating a BESS with an EVCS, it requires consulting with reliable installers and adhering to local laws and safety standards.

In conclusion, homeowners will benefit from a number of advantages from combining a BESS and an EVCS. These include of centralized control and management, grid integration, load management, energy optimization, and backup power capabilities. Users can raise their energy resilience, lower their energy expenses, and increase their energy efficiency by integrating these two systems. Consult with reliable installers, and adhere to local laws and safety standards guarantees to ensure correct installation, electrical compatibility, and optimal system performance.

10.5 CONCLUSIONS

This chapter's examination of battery technology highlights how important it is becoming to household energy systems due to the quick rise of renewable energy sources and rising need for energy storage. The chapter starts with an overview of several battery technologies and then extensively discusses the problems that arise, such as poor capacity, excessive costs, and environmental concerns at every stage of manufacture and disposal. The chapter sets the stage for a thorough analysis of possible domestic applications by highlighting the critical need for increased safety, enhanced performance, and environmentally friendly production methods. Emphasizing the vital role that batteries play in load shifting, peak shaving, grid resilience, and boosting renewable energy self-consumption, the conversation also covers the battery's integration with solar panels, smart home systems, and infrastructure for electric car charging. Here, the report outlines the advantages as well as issues to consider for homeowners who have considered these advances.

The emphasis on current advancements and new trends highlights their capacity to get over current limitations and present innovative alternatives for home energy systems. The final section highlights how crucial it is for stakeholders, policymakers, and researchers to work together and conduct ongoing research in order to fully realize the potential of battery technology for home energy systems. The chapter also envisions a more resilient and sustainable future and presents a picture of revolutionary change in home energy management. It offers important insights into the changing environment of greener and more efficient home energy systems by carefully negotiating opportunities and challenges. This thorough investigation not only recognizes the existing complexity but also highlights battery technology as the primary driver behind developing a home energy future that is robust and ecologically aware.

The chapter also discusses how combining BESS with EVCS can be an innovative strategy that offers a number of benefits to homeowners. A multitude of advantages are provided by the synergistic integration of various systems, including improved grid interface, backup power capability, centralized control and management, energy optimization, and effective load management. This combination not only promotes energy efficiency but also gives homeowners a chance to reduce energy expenses and strengthen their energy resilience. Users can navigate through dynamic energy demands thanks to the seamless coordination between BESS and EVCS, which helps create a more affordable and sustainable home energy ecology. Moreover, the incorporation of backup power features guarantees a dependable energy supply in the event of unplanned disruptions, enhancing the total energy reliability for households. It is crucial to place a high priority on accurate installation, guaranteeing electrical compatibility, and maximizing system performance for individuals thinking about integrating BESS and EVCS. To fully utilize these integrated systems, it is imperative to adhere to local rules, follow safety protocols, and consult with reputable installers. By taking a comprehensive approach, possible hazards are avoided and the advantages of this cutting-edge energy solution are optimized. Overall, using BESS and EVCS together is an innovative approach for homes to manage their energy in a resilient and sustainable way. As technology continues to advance, embracing such integrated solutions becomes pivotal for homeowners seeking to navigate the complexities of contemporary energy challenges.

11 Electric Vehicles as an Active Energy Storage System

*Youssef Achour, Anisa Emrani,
Asmae El Moukrini, and Asmae Berrada*

11.1 INTRODUCTION

Automobiles enable people to work, travel, transport goods, and provide countless other services. There are over 1.2 billion vehicles worldwide; indeed, the automobile sector is one of the largest and fastest-growing contributors to the global economy. Their huge impact justifies the intense interest of researchers and developers in advancing this field [1]. Given that transportation facilities represent a large contributor to greenhouse gas emissions, the last few years have seen a significant shift towards electric vehicles (EVs). EVs are a prominent solution to the air pollution issue, according to the International Energy Agency (IEA) [2]. However, in order to facilitate a sufficient uptake of this new generation of electric vehicles, a robust and compact infrastructure with supportive social and political policies is needed [3]. The EU20 nations and Brazil, along with SAR, continue to implement favorable tax policies for electric transportation. London has introduced fees for accessing in-city roads, with exemptions for EVs [4]. Figure 11.1 demonstrates some of these aspects regarding incentives, policy regulations, and future goals [5].

There are numerous studies on this subject. Keshavarzmohammadian *et al.* [6] studied emissions from light electric vehicles in the USA. Their findings suggest that by 2050, a 47 percent battery electric (BEV) fleet could result in a 36 percent reduction in greenhouse gas emissions in the transportation sector. Navas-Anguita *et al.* [7] studied the carbon footprint of alternative fuel production technologies in Spain. Wind energy and solar technologies would have a significant role in meeting the EVs' energy requirements. Researchers were interested in developing accurate models to accommodate the energy usage of EVs. A traffic flow theory and locomotion mechanics were presented in 2020 by a team of researchers to get an accurate calculation regarding energy consumption [8]. Several comparative studies were conducted to evaluate the performance of different EV types. Sheng *et al.* [9] compared the insertion of electric vehicles in Australian and New Zealand markets. They concluded that BEVs outperformed other types in terms of energy consumption in these countries. Cradle-to-grave life cycle assessment of different types of vehicles was assessed by Karaaslan *et al.* in 2018 [10]. Their findings testified that battery electric vehicles had the lowest greenhouse gas emissions and energy consumption, despite higher emissions in the manufacturing phase.

DOI: 10.1201/9781003441236-11

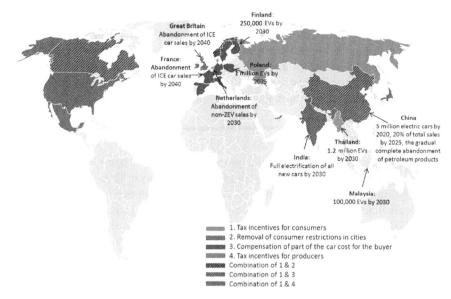

FIGURE 11.1   Electric vehicle policies and goals in different regions across the globe.

This chapter's goal is to present a scientific outlook on the potential of EV as an active energy storage system. The concept of vehicle-to-grid (V2G) has presented significant potential for regulating and responding to the grid's energy demand. This could encompass discussions about the efficiency of energy storage in EV batteries, advancements in battery technologies, the impact of EVs on the electricity grid, and related scientific considerations. The aim is to get valuable insights about the topic of energy storage.

11.2   RESIDENTIAL POWER SYSTEM

11.2.1   VEHICLE-TO-GRID

Electric grid infrastructure research stands out as a very prominent field that requires innovation and exploration. Local energy production, energy management, and microgrid connections are key components in creating resilient and flexible smart cities [11]. A literature review offers a variety of studies on this topic. Energy resource scheduling for smart grids was established in [12]. The authors in the latter paper have studied intensive usage of distributed generation alongside gridable vehicles. The demand response was also a key factor in several studies.

Two proposed applications were considered in [13], with the first addressing a gas turbine power plant capable of handling intermittent generation. The second application proposes a demand-side management (DSM) program for residential customers. They included plug-in hybrid electric vehicle (PHEV) charging rescheduling and direct load control (DLC) for air conditioning (AC) and electric water heating (EWH) units. V2G is highly efficient for peak shaving (PS) (powering the grid during high demand) and valley filling (charging at night during low demand) [14]. In V2G applications, ensuring dependable electronic detection and establishing a robust connection between the EV and the load is of utmost importance. The EV is connected to the source on a regular basis; therefore, the following goals can be achieved: charging the EV battery, storing the power for later use or to power EV, and reinjecting the stored power back into the grid to respond to the demand.

11.2.2 BATTERIES

Batteries are a mature electrical power storage technology. Their main components include an anode (negative), a cathode (positive), and a catalyst. There are several commercially available types of batteries, such as primary batteries (non-rechargeable), primary alkaline batteries, primary lithium batteries, flow air batteries, and secondary batteries [15]. Table 11.1 represents different manufacturers of electrical vehicles alongside the batteries used in each vehicle model. The state of charge of a battery, expressed in Equation (11.1), is a key parameter to evaluate its performance:

$$SOC = \frac{\int_{T_0}^{t} I(t)dt}{Q_{rated}} \# \tag{11.1}$$

TABLE 11.1
Batteries used in EVs from various manufacturers

Company	Country	Vehicle model	EV type	Battery type	Capacity (kWh)
Toyota	Japan	Prius PHV (Prime)	PHEV	Li-ion	8.8
		Prius (fourth generation)	HEV	Ni-MH	1.31
		Prius (fourth generation)	HEV	Li-ion	0.75
		Aqua (Prius C)	HEV	Ni-MH	0.94
Nissan		Leaf	BEV	Li-ion	30
Honda		Accord Hybrid	HEV	Li-ion	1.3
		Fit (Jazz) Hybrid	HEV	Li-ion	0.86
Mitsubishi		i-MiEV	BEV	Li-ion	16
		Outlander	PHEV	Li-ion	12
BMW	Germany	i3	BEV	Li-ion	33
		x5 xDrive40e	PHEV	Li-ion	9
Mercedes-Benz		B250e	HEV	Li-ion	28
Audi		A3 Sportback e-tron	PHEV	Li-ion	8.3
Volkswagen		e-Golf	BEV	Li-ion	35.8
Volvo	Sweden	XC90 T8	PHEV	Li-ion	9
Fiat	Italy	500e	BEV	Li-ion	24
Tesla	USA	Model S	BEV	Li-ion	60–100
General Motors		Chevrolet Volt	PHEV	Li-ion	18.4
Ford		C-MAX Energi	PHEV	Li-ion	7.6
Hyundai	South Korea	Sonata Hybrid	HEV	Li-polymer	1.6

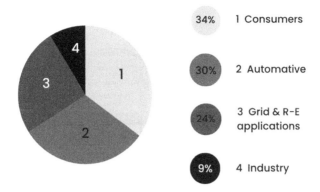

FIGURE 11.2 Market segments for Li-ion batteries.

where SOC represents the state of charge of the battery, I(t) is the current flow in and out of the battery, Q_{rated} is the rated capacity of the battery, t indicates time variation, and T_0 is the initial time.

Transparency Market Research released a report estimating the global Li-ion battery market, which was valued at $1.7 billion in 2012. Its growth surpassed the compound annual growth rate (CAGR) by 14.4 percent, which was about $33.11 billion in 2019 [16]. The market sector for Li-ion batteries is depicted in Figure 11.2.

11.2.3 ELECTRIC VEHICLES

Electric vehicles represent a large share of the upcoming rise in automobile manufacturing. Important parts of an electric car include:

- *Low-voltage auxiliary battery*. Before turning on the traction battery, low-voltage auxiliary battery starts the car and powers accessories.
- *High-voltage battery pack*. This latter provides additional power to the electric traction motor by storing energy from regenerative braking.
- *DC-DC converter*. This component converts the traction battery's higher-voltage DC power into lower-voltage DC power for auxiliary battery recharging and accessory usage.

Table 11.1 sheds light on the main EV contributors, types, and batteries used, while Figure 11.3 displays the different EVs that exist in today's market, which fall into one of four categories [17, 18]:

- *PHEV: Plug-in hybrid vehicles*. This EV type combines an internal combustion engine with an electric propulsion system. PHEVs are designed to operate in both all-electric mode and hybrid mode. They can travel short distances while being in full electric mode. When the battery is nearly depleted, the internal combustion engine engages. These specific operating conditions reduce both

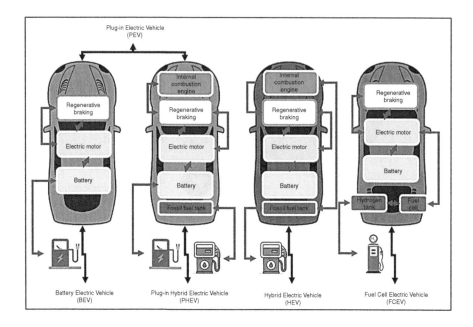

FIGURE 11.3   The four types of EVs.

fuel consumption and emissions rates compared to traditional internal combustion vehicles.

- *BEV: Battery electric vehicles*. The EVs in this category are dependent solely on the integrated battery. The engine is fully electrical and derives the necessary power from the rechargeable battery. Typically, recurrent braking actions can be utilized to recapture electrical energy for the engine, taking advantage of the elevated heat in the brake disks. BEVs employ power electronics and electric motors to achieve the electric-to-mechanic conversion, propelling the vehicle. There are no emissions during the operating period associated with this type of EV.

- *HEV: Hybrid electric vehicles*. Similar to plug-in hybrid vehicles, hybrid vehicles combine, on their own, an internal combustion engine with an electric system. The difference is found in the charging mode, where the HEVs do not require external charging. Instead, they rely on regenerative braking and engine-driven generators as self-charging mechanisms. HEVs display high efficiency, especially during low-speed drives and during acceleration, where the electric system assists the engine.

- *FCEV: Fuel cell electric vehicles*. This type of vehicle operates on hydrogen fuel. The integrated fuel cell combines hydrogen and oxygen atoms to form water molecules and electric current at the same time. They are expected to have a strong presence in the global EV market. Their water-only emissions make them the cleanest means of transportation. Further research is necessary to enhance the efficiency of the fuel cell device and ensure adequate integration. The main components of an FCEV can be summarized as follows. A stack of membrane electrodes used in fuel cells to produce energy from hydrogen and oxygen. A tank for hydrogen and gasoline filler that holds and releases hydrogen for the fuel cell as required. The power electronics controller regulates the electrical energy flow between the fuel cell and battery in addition to controlling the electric motor's torque and speed. Additionally, it controls the cooling system, or thermal system, which maintains the electric motor, fuel cell, power electronics, and other associated parts at optimal operating temperatures.

11.3 CASE STUDY

Figure 11.4 illustrates a schematic of an overall residential grid connected to an EV. Both grid-to-vehicle and vehicle-to-grid functionalities signify bidirectional power flow. V2G allows EVs to act as a power source, injecting power back into the grid during additional demand. G2V enables charging EVs from the grid, utilizing available energy resources efficiently. This back and forth of power is what guarantees flexible energy usage.

11.3.1 The Hybrid Energy Generation Model

Solar Cell Electric Equivalent Model
Figure 11.5 shows the electrical equivalent of a solar cell; it displays a simplified representation with one diode.

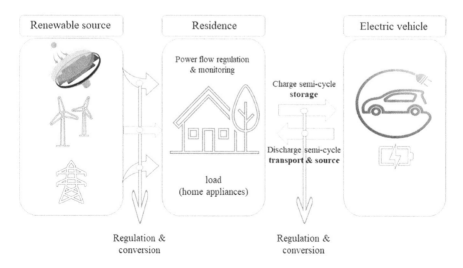

FIGURE 11.4 Residential grid with renewable energy generation and EV technology.

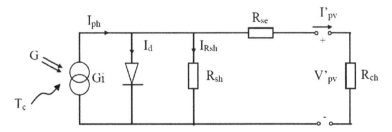

FIGURE 11.5   Electrical equivalent model of a solar cell.

The cell current or the generated current is expressed as in Equation (11.2):

$$I_{cell} = I_{ph} - I_{01} \cdot \left[\exp\left(\frac{V_{cell} + R_s . I_{cell}}{n_1 . V_{th}} \right) - 1 \right] - \frac{\left(V_{cell} + R_s . I_{cell} \right)}{R_{sh}} \qquad (11.2)$$

I_{ph} concerns the temperature and the radiation distribution and influence; it is expressed as shown in Equation (11.3):

$$I_{ph} = \left[I_{sc} + K_i \left(T - T_{ref} \right) \right] \left(\frac{G}{G_0} \right) \qquad (11.3)$$

K is the Boltzmann constant (1.38×10^{-23} J/K). I_s represents the saturation current; it is expressed as in Equation (11.4):

$$I_s = I_{rs} \cdot \left(\frac{T}{T_{ref}} \right)^3 \cdot \left[\exp\left(\frac{1}{T_{ref}} - \frac{1}{T} \right) \right] \cdot \frac{E_g . q}{K.A} \qquad (11.4)$$

I_{rs} is the reverse saturation current of the cell at the reference temperature T_{ref} (25 °C). It is expressed by Equation (11.5):

$$I_{rs} = \frac{I_{sc}}{\exp\left(\frac{q.V_{sc}}{A.K.N.N_s} \right) - 1} \qquad (11.5)$$

where E_g is the gap band energy, I_{sc} is short current circuit, and q is the electron charge (1.6×10^{-19} C). The mathematical model of the solar cell is expressed as in Equation (11.6):

$$I = N_p \cdot \left(\left(I_{ph} - I_s \left(\exp\left(\frac{q.\left(V + R_s . I \right)}{N_s . A.K.T} \right) \right) \right) - 1 \right) - I_{sh} \qquad (11.6)$$

N_s and N_p are the number of modules connected in series and in parallel, respectively.

Wind Turbine Mathematical Modeling

The mechanical power P_{wind} extracted by the rotor blades in watts is the difference between the upstream and the downstream wind powers. It is displayed in Equation (11.7) [19]:

$$P_{wind} = \frac{1}{2} \rho.A.v_w \left(v_u^2 - v_d^2 \right) \tag{11.7}$$

where A is the area of wind captivity and ρ is the density of air. The parameters v_w, v_u, and v_d represent the wind velocity, the upstream wind velocity at the entrance of the rotor blades, and the downstream wind velocity at the exit of the rotor blades in m/s, respectively.

The calculation of the electric power generated by the wind turbine implies identifying the ratio of wind speed v_d downstream to wind speed v_u upstream of the turbine λ, as in Equation (11.8). This ratio is later used to identify the power coefficient C_p, as shown in Equation (11.9); it represents the efficiency in capturing available wind power:

$$\lambda = \frac{\text{blade tip spead}}{\text{wind speed}} = \frac{v_d}{v_u} \tag{11.8}$$

$$C_p = \frac{(\lambda + 1)(1 - \lambda^2)}{2} \tag{11.9}$$

Finally, the electrical power, Equation (11.10), can be extrapolated from Equations (11.7–11.9):

$$P_{elec} = C_p \times P_{wind} \tag{11.10}$$

11.3.2 RESIDENTIAL GRID MODELING

The residential grid under investigation consists of a PV array combined with a small wind turbine to generate renewable energy, a vehicle-to-grid (V2G) system managing battery charging and grid regulation, and a load. This latter serves a small community (household). The model inputs include solar irradiation and temperature variation during a 24-hour period, as well as wind speed profile over the same period. The model was developed using Python. Specific libraries are required including Matplotlib, NumPy, Gekko, and FcPy, as they provide mathematical outlook, and visualization solutions. The load comprises typical residential appliances (e.g., lights, washing machine, cooking devices, and ventilation system). This standard load follows a typical consumption pattern for an eight-person household. A 30 kWh daily average load was chosen for this simulation. Figure 11.6 is a flowchart of the energy generation and distribution in the residential grid. The simulation spans 24

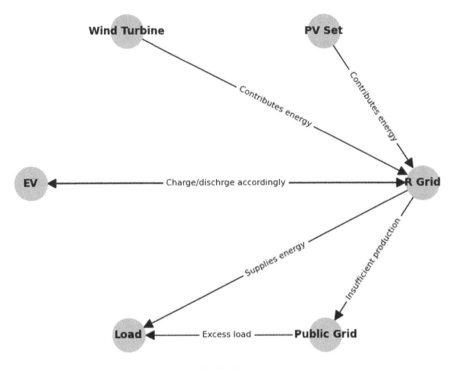

FIGURE 11.6   Energy production and distribution.

hours following a normal distribution of solar irradiation, peaking around noon. Wind conditions vary considerably throughout the day. The residential load mimics typical household consumption, reaching a peak in the evening and declining overnight.

The location chosen for this simulation is Rabat, Morocco. With a latitude of 33.97 and a longitude of 6.84, this city is famous for its Mediterranean climate type, according to the Köppen Climate Classification [20]. To acquire valuable input on residential grid energy usage and distribution, data such as temperature variation and irradiation are necessary. The model input data were obtained from PVGIS platform [21] and the Global Wind Atlas platform [22]. Figure 11.7 demonstrates the variation of these parameters in an average monthly step. The temperature varies typically between 25 °C and 14 °C. Strong irradiation is observed throughout the year, especially in the months between March and September; it reaches its peak during the month of July. Wind speed varies from 5 m/s to 10 m/s at an altitude of 10 m. The data show a suitable wind speed for micro to small wind turbine installations. The speed can introduce wind power without damaging the devices. Table 11.2 presents various characteristics of the components of the residential system under study [23, 24]. The costs of the investigated vehicles, along with hydrogen prices in 2023, are provided in Table 11.3.

- *Scenario 1.* The selected EV for the first scenario is a BEV. The V2G system's inclusion serves two main roles, the first of which concerns charging the EV's

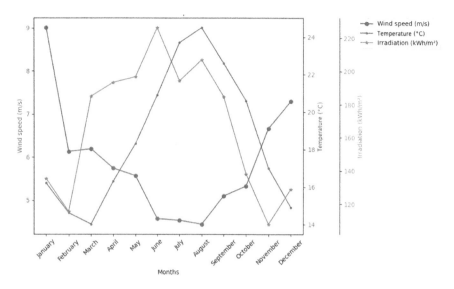

FIGURE 11.7   Wind speed, irradiation, and temperature monthly average variation.

TABLE 11.2
Power and financial parameters of different subsystems

Subsystem	Wind turbine	PV array	Public grid
Rated power or capacity	$3\ kW_p$	$2.5\ kW_p$	On demand
Type	Permanent magnet	Monocrystalline silicon	—
Life span	27 years		—
CapEx	2,57 $/Wp	0.3 $/Wp	0.09 $/kWh
OpEx (% of CapEx)	1%	2%	—

TABLE 11.3
Vehicles and hydrogen costs

Application	Targeted price	Unit	Reference
FCEV	3551	$/kW	[25]
Hydrogen supply	4.34	$/kg	[23]
EV	3318.2	$/kW	[25]

battery for later usage and/or energy storage, and the second of which is about managing grid regulation and utilizing available power to stabilize the grid in case of disruptions. The battery is lithium-ion-based with a 24 V rated voltage and 11 kW rated power. The self-discharge of the battery is about 1 percent per month. The temperature sensitivity is an important factor that has been considered given its major effect on the battery.

- *Scenario 2.* The second scenario make use of an FCEV in the V2G system. The FCEV serves the same role as the BEV described in the first scenario. FCEV is hydrogen fuel-based. The electric charge passes through a separate electrolyzer to produce the necessary hydrogen. The produced hydrogen is later stored in appropriate tanks in its gaseous state. The rated power of both the electrolyzer and the fuel cell is equal to 12.25 kW. The integrated polymer exchange membrane fuel cell (PEMFC) has 378 cells. Each sell has an active area of about 300 cm² and a nominal voltage of 0.6 V. The maximum current supported by a cell is in the range of 540 A. To establish a smooth operation, the total active surface of the FCEV is 11.34 m².

11.4   RESULTS AND DISCUSSION

The hybrid renewable energy system studied in this chapter includes an energy dispatching system to monitor and respond to energy demand accordingly. A 24-hour simulation (a full day) was run to visualize the interactions between the different subsystems energy-wise. The chosen day was in the month of July. The refrigerator, as an example, has a rated power of 0.8 kW and works all day long. The permutation between charging and discharging is accomplished by the monitoring system. Figure 11.8 shows the variation of the battery's SOC alongside the voltage. On the one hand, granting a full uninterrupted charge to the battery while also not exhausting it with high voltages helps with the battery longevity and reduces the risks of degradation. On the other hand, not allowing a deep discharge where the battery is almost depleted entirely before recharging has positive effects. Furthermore, the deep discharge might accelerate the deterioration of the EV's battery. The semi-cycles of charging and discharging take 24 hours to complete. The manufacturers of EVs are under obligation to guarantee at least eight years of usage for their batteries. If kept in good condition, a battery can last up to ten years without the need to change it.

Figure 11.9 displays the energy generation of the different subsystems as well as the load demand. The vehicle is supposed to be stationary and connected to the V2G system. Figure 11.10 demonstrates the energy distribution where the BEV is not connected to the V2G system from 8 a.m. to noon and from 2 p.m. to 6 p.m. (a typical working day). The public grid intervenes in this case to respond to the energy shortage from the renewable energy sources. The noticeable energy consumption follows two peaks: one in the morning, when everyone wakes up and heads to school or work; and the second in the evening, when everyone returns home. The typical household with an average 30 kWh energy demand presents predictable repeatable behavior throughout the days. That is the reason behind choosing 24-hour simulation span. The battery is noticeably in charging mode during the middle of the day; and it reaches a full SOC at approximately 3 p.m. in the first case. The energy generated by the small wind turbine and the PV set overcovers the load demand. Therefore, the extra energy is directed to charge the battery. In the evening peak, when the PV system stops generating energy, the battery takes over and switches to discharging mode. The first case enables proper energy distribution without the need to utilize or

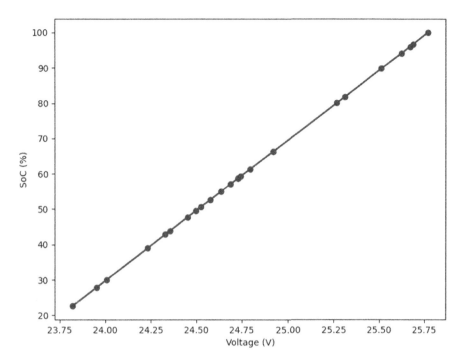

FIGURE 11.8   State of charge variation with battery's voltage.

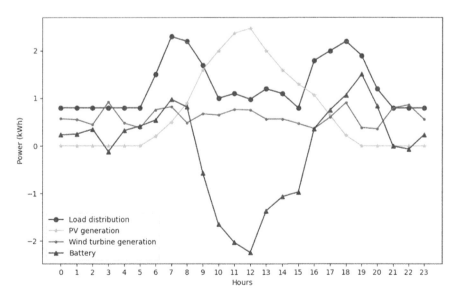

FIGURE 11.9   Energy generation and demand for the residential grid (first case).

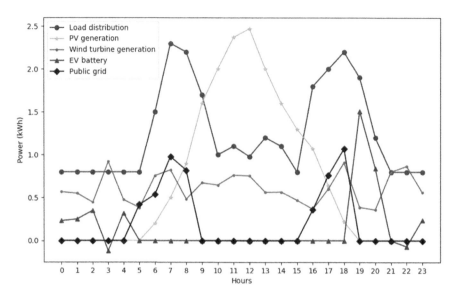

FIGURE 11.10   Energy generation and demand for the residential grid (second case).

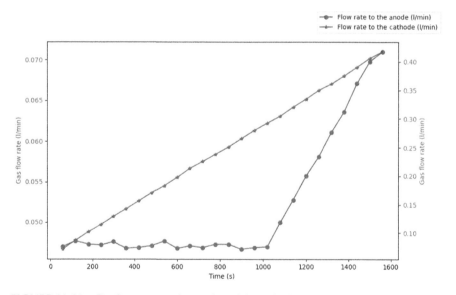

FIGURE 11.11   Gas flow rates to the anode and the cathode.

integrate the public grid. Load-shifting in the studied system ensures grid stability. By constantly evaluating energy generation (input) and energy consumption (output), it helps avert strain or possible blackouts during moments of high demand.

The temperature of the cell in this configuration is around 80 °C. Both the anode and cathode have a temperature of 40 °C. Figure 11.11 presents the gas flow

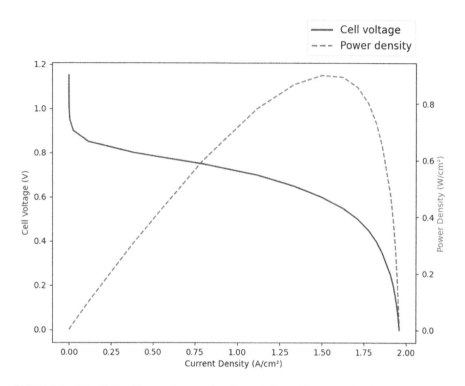

FIGURE 11.12   Cell voltage and power density variation with current density.

introduced to the anode and the cathode at their respective temperatures (40 °C). A well-measured flow rate results in favorable results in terms of electricity production. Subsequently, at the cathode, oxygen molecules (O_2) from the air are supplied. They combine with electrons and protons which travel through the electrolyte from the anode. This leads to the reduction of oxygen into hydroxide ions (OH^-) via the oxygen reduction reaction (ORR). The current density is observed in Figure 11.12. The peak is about 1.55 A/cm^2.

Similar to the BEV case, Figure 11.13 displays the energy generation and distribution among the various components of the residential grid. The vehicle investigated in this case is an FCEV. The time response of the fuel cell is much lower than that of BEV. The downside of this configuration is the energy waste that will occur if an energy storage system is not included. Hydrogen (being the fuel for the vehicle) is another challenge. Hydrogen production was not discussed in this chapter. It is shown that the public grid is more utilized in this case for the provision of energy.

The energy savings calculations serve as a crucial metric for assessing the feasibility of various approaches. The computations primarily accounted for residential grid energy costs and did not encompass additional factors like transportation. The results illustrated in Figure 11.14 show that residences leveraging V2G integration with BEVs yield higher energy savings, amounting to $9362.25. On the other hand, the energy savings for the project utilizing FCEVs amount to $6405.75. Considering

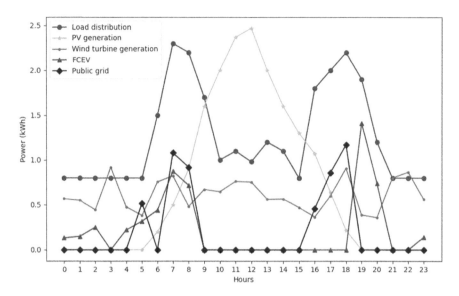

FIGURE 11.13   Energy generation and demand for the residential grid.

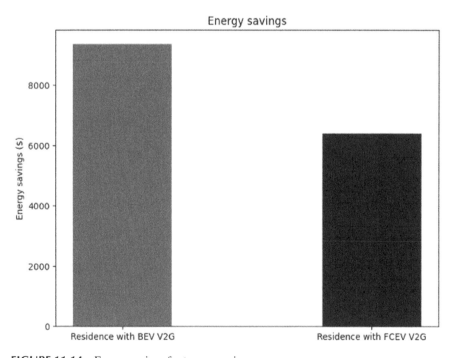

FIGURE 11.14   Energy savings for two scenarios.

energy aspects, implementing V2G systems to fulfill energy demands has been proven to be the more advantageous strategy. The system's broader applications, particularly in transportation and the pursuit of nearly attaining energy independence from the public grid, V2G systems emerge as an excellent alternative.

11.5   CONCLUSION

This chapter falls within the discourse on innovative means to store energy, especially in residential use. A synopsis of BEVs, PHEVs, HEVs, and FCEVs is presented. The market shares, large manufacturers, and future aspirations are discussed. The performance of the EV's Li-ion battery, and its integration as an active energy storage system are examined. A 24-hour simulation of the grid's energy distribution is conducted. The permutation of the charging and discharging semi-cycles is covered. Load-shifting and the hybridization of energy generation modes have proven to be efficient in terms of ensuring the grid's stability. Further optimization of the system is required to find the ideal solution for energy storage. A comparison between the performance of the other EVs in this scenario is highly recommended. V2G systems, however, become a strong alternative when considering system uses that go beyond energy delivery, such as mobility and the pursuit of increased energy autonomy from the public grid. Peak demands present a significant challenge to the grid. The energy dispatching system requires continuous monitoring to avoid power outages. Smart grids offer an improved version of the conventional grid by enhancing communication and including the hybridization of energy resources. Nevertheless, renewable energy means they utilize pose a much lesser risk to the environment than conventional grid. Findings indicate that the residential grid with a BEV in conjunction with V2G integration saves more energy, totaling $9362.25. Conversely, the project employing FCEVs results in energy savings of $6405.75. The future vision is to create a coordination between smart microgrids to make a large smart grid, where energy dispatching is used locally and sold to other microgrids in need, and so forth.

This chapter opens the door to several perspectives and future studies. Both configurations have a noteworthy effect on the climate. CO_2 emissions can be reduced significantly. Studying the environmental impact and the life cycle assessment is vital. In addition, the prices are still high, which makes the application of these technologies not yet at everyone's disposal. Discovering new materials and minimizing the overall cost should also be tackled.

NOMENCLATURE

A	Area of wind captivity (m²)
CapEx	Capital expenditure of the project ($)
C_p	Power coefficient
E	Electrical energy (kWh)
EP	Energy production
η	PV system efficiency (%)

H_t	Solar radiation (kWh/m²)
Impp	Current at maximum power point
Isc	Short-circuit current (A)
m	Flow rate (l/min)
M	Total mass of hydrogen produced (kg)
n	Molecular flow rate (kmol/h)
OpEx	Operational and maintenance expenditure ($)
r	Discount rate (%)
R_d	Degradation rate (%)
T	Temperature (°C)
v_d	Wind speed downstream (m/s)
v_u	Wind speed upstream (m/s)
v_w	Wind speed (m/s)
λ	Ratio of wind speed upstream to downstream

Abbreviations

BEV	Battery electric vehicle
EV	Electric vehicle
FCEV	Fuel cell electric vehicle
G2V	Grid-to-vehicle
HEV	Hybrid electric vehicle
IRR	Internal rate of return
ORR	Oxygen reduction rate
PHEV	Plug-in hybrid electric vehicle
SOC	State of charge
V2G	Vehicle-to-grid

REFERENCES

[1] T. Muneer, M. Kolhe, and Doyle, A. 2017. *Electric vehicles: Prospects and challenges*. Elsevier.

[2] IEA, *Global EV Outlook 2022*. Paris: IEA, www.iea.org/reports/global-ev-outlook-2023

[3] F. Knobloch *et al.*, "Net emission reductions from electric cars and heat pumps in 59 world regions over time," *Nat. Sustain.*, vol. 3, no. 6, pp. 437–447, Mar. 2020, doi: 10.1038/s41893-020-0488-7

[4] IEA, *Global EV Outlook 2023: Analysis*. Paris: IEA, www.iea.org/reports/global-ev-outlook-2023

[5] N. O. Kapustin and D. A. Grushevenko, "Long-term electric vehicles outlook and their potential impact on electric grid," *Energy Policy*, vol. 137, p. 111103, Feb. 2020, doi: 10.1016/j.enpol.2019.111103

[6] A. Keshavarzmohammadian, D. K. Henze, and J. B. Milford, "Emission impacts of electric vehicles in the US transportation sector following optimistic cost and efficiency projections," *Environ. Sci. Technol.*, vol. 51, no. 12, pp. 6665–6673, Jun. 2017, doi: 10.1021/acs.est.6b04801

[7] Z. Navas-Anguita, D. García-Gusano, and D. Iribarren, "Long-term production technology mix of alternative fuels for road transport: A focus on Spain," *Energy Convers. Manag.*, vol. 226, p. 113498, Dec. 2020, doi: 10.1016/j.enconman.2020.113498

[8] A. I. Croce, G. Musolino, C. Rindone, and A. Vitetta, "Energy consumption of electric vehicles: Models' estimation using big data (FCD)," *Transp. Res. Procedia*, vol. 47, pp. 211–218, 2020, doi: 10.1016/j.trpro.2020.03.091

[9] M. S. Sheng, A. V. Sreenivasan, B. Sharp, and B. Du, "Well-to-wheel analysis of greenhouse gas emissions and energy consumption for electric vehicles: A comparative study in Oceania," *Energy Policy*, vol. 158, p. 112552, Nov. 2021, doi: 10.1016/j.enpol.2021.112552

[10] E. Karaaslan, Y. Zhao, and O. Tatari, "Comparative life cycle assessment of sport utility vehicles with different fuel options," *Int. J. Life Cycle Assess.*, vol. 23, no. 2, pp. 333–347, Feb. 2018, doi: 10.1007/s11367-017-1315-x

[11] F. Y. Melhem, N. Moubayed, and O. Grunder, "Residential energy management in smart grid considering renewable energy sources and vehicle-to-grid integration," in *2016 IEEE Electrical Power and Energy Conference (EPEC)*, Oct. 2016, pp. 1–6, doi: 10.1109/EPEC.2016.7771746

[12] T. Sousa, H. Morais, J. Soares, and Z. Vale, "Day-ahead resource scheduling in smart grids considering Vehicle-to-Grid and network constraints," *Appl. Energy*, vol. 96, pp. 183–193, Aug. 2012, doi: 10.1016/j.apenergy.2012.01.053

[13] R. Roche, "Agent-based architectures and algorithms for energy management in smart gribs: Application to smart power generation and residential demand response algorithmes et architectures multi-agents pour la Gestion de l ' Énergie dans les Réseaux Électriques Int," 2012, [Online]. Available: https://tel.archives-ouvertes.fr/tel-00864268/document

[14] T. M. Letcher (ed.), *Future Energy: Improved, Sustainable and Clean Options for Our Planet.* Elsevier, 2014. doi: 10.1016/C2012-0-07119-0

[15] M. A. Abdelkareem *et al.*, "Environmental aspects of batteries," *Sustain. Horizons*, vol. 8, p. 100074, Dec. 2023, doi: 10.1016/J.HORIZ.2023.100074

[16] "Rechargeable Poly Lithium Ion Batteries Market – Segment, Forecast 2024." www.transparencymarketresearch.com/rechargeable-poly-lithiumion-batteries-market.html (accessed Nov. 22, 2023).

[17] "The ICE Age is Over: Why Battery Cars will Beat Hybrids and Fuel Cells." https://thedriven.io/2018/11/14/the-ice-age-is-over-why-battery-cars-will-beat-hybrids-and-fuel-cells/ (accessed Nov. 13, 2023).

[18] R. R. Kumar and K. Alok, "Adoption of electric vehicle: A literature review and prospects for sustainability," *J. Clean. Prod.*, vol. 253, p. 119911, Apr. 2020, doi: 10.1016/j.jclepro.2019.119911

[19] K. E. Johnson, "Adaptive Torque Control of Variable Speed Wind Turbines," Golden, CO (United States), Aug. 2004. doi: 10.2172/15008864

[20] "Köppen Climate Classification System." https://education.nationalgeographic.org/resource/koppen-climate-classification-system/ (accessed Nov. 22, 2023).

[21] "JRC Photovoltaic Geographical Information System (PVGIS) – European Commission." https://re.jrc.ec.europa.eu/pvg_tools/en/#api_5.1 (accessed Nov. 20, 2023).

[22] "Global Wind Atlas." https://globalwindatlas.info/fr (accessed Nov. 20, 2023).

[23] Y. Achour, A. Berrada, A. Arechkik, and R. El Mrabet, "Techno-economic assessment of hydrogen production from three different solar photovoltaic technologies," *Int. J. Hydrogen Energy*, vol. 48, pp. 32261–32276, May 2023, doi:10.1016/j.ijhydene.2023.05.017

[24] IRENA, "Renewable Power Generation Costs in 2021 – Chart data," *Renew. Power Gener. Costs 2021*, 2022, [Online]. Available: www.irena.org/publications/2022/Jul/Renewable-Power-Generation-Costs-in-2021 (accessed Apr. 05, 2023)

[25] V. T. Office, "2022 Incremental Purchase Cost Methodology and Results for Clean Vehicles," Dec., 2022, [Online]. Available: www.energy.gov/eere/vehicles/21st-century-truck-partnership

12 Virtual Power Plants with Battery-Integrated Residential Systems

Mohammad Reza Fallahzadeh, Ali Shayegan-Rad, and Ali Zangeneh

12.1 INTRODUCTION

Power systems benefit from decentralized power generation (DPGs) based on renewable resources as they face the challenges of power market liberalization and environmental protection [1]. However, these renewable resources (RRs) have no ability to adjust their output power to the varying conditions of their natural sources. This challenge makes their power outputs fluctuate randomly [2]. The low and uncertain generation of RERs makes it difficult for these resources to take part in the energy market [3]. To address this issue, the virtual power plant (VPP) has been introduced as a way to combine and coordinate DPGs (either renewable or conventional generation) with a sufficient capacity to enable them to participate in the energy market and also to enhance the technical performance of its distribution system by applying proper management of DPGs [2, 4, 5].

The idea and structure of the VPP was proposed by Awerbuch and Preston in 1997 for the first time [6]. Pudjianto et al. [7] explored the notion of this framework to enable cost-effective integration of existing DPGs into the power system. The VPP can also provide benefits such as reducing the operational costs, increasing the reliability and resilience, and supporting the grid stability and security [8]. A VPP can consist of various kinds of generation and storage units that work together like a single unit conventional power plant to participate in the power market with a certain profit in each hour [3]. The VPP plays a role like a unified player in the power market, even though it may include different DPGs [8].

Pudjianto's study [9] proposed a new frame to facilitate the involvement of DPGs in the electricity market. It showed that the technical and economical features of each DPGs have been improved using the VPP notion. A accidental optimal modeling of VPP considering the correlations between incontrollable decentralized generation units such as wind base turbine and solar cell generations is presented in [10]. The output power probability of independent wind and solar cell resources are modeled by empirical distribution functions, and their correlation factor model is efficiently determined using the Frank copula function. Furthermore, some VPPs that use non-fossil technologies to produce electricity have installed small or large number of batteries to store energy and release it when distribution system required or during peak hours in which optimizes VPP operation.

DOI: 10.1201/9781003441236-12

In this chapter, small batteries are considered as a battery energy storage system (BESS) to generate power and participate in energy market in order to make more profit for the VPP. Gonçalves et al. [11] used batteries similar to a single storage entity to introduce a strategy for managing energy in peak and none peak hour, due to sever use of energy resources. With the emergence of electric vehicles (EVs), parking lots which use as EVs aggregator can model as a virtual DPG to take part in the power market [12]. An optimal scheduling of energy which used several distributed generations and EVs is discussed in [13], which performs an intensive analysis of the DPG costs and capacities, battery and gasoline prices. These studies demonstrate the potential and challenges of VPPs in integrating BESS sources beside various DPGs into the power system and market. A novel model for programing a single BESS in the day-ahead energy, spinning reserve and RR markets is proposed in [14]. Some of these papers have also considered integration of small residential BESS into a VPP to improve the technical and economical characteristic of the distribution network by participating in the power market. Pandžić et al. [15] proposed a VPP model which use different periodic energy resources, GT, and small BESS to take part in day-ahead energy markets to maximize its expected profit. This chapter presents a novel model for optimal scheduling of a VPP consisting of various BESS to provide energy to power market.

This chapter proposes an optimal strategy for making decision of VPPs to take part in the day-ahead market. The VPP consists of wind turbine (WT), small BESS, and gas turbine (GT). The main points of the chapter can be summarized as follows:

1. The designed model includes several types of energy generation (GT, WT, and BESS) in the day-ahead markets;
2. To aggregate a large number of residential battery storage system, an entity is defined to aggregate and manage the charging and discharging of these storage facilities;
3. The VPP sets up incentive contracts with BESS owners to use their energy and act as a DPG; and
4. The reprimand factor for CO_2 emissions is considered in the GT cost function.

The rest of the chapter is structured as follows. Section 12.2 elucidates the VPP model and presents a formulation for VPP decision strategy to take part in energy market. Section 12.3 presents the case studies to validate the proposed model and also shows the results obtained. Finally, the conclusions are presented in Section 12.4.

12.2 THE VPP OPTIMAL SCHEDULING MODEL

The VPP model under consideration includes a GT, a WT [16], and a large number of small BESS assumed to a single and with high capacity by means of an aggregator. The VPP purchases energy to engage in the day-ahead market [17–19]. Small residential BESS are modeled as electrical storage sources, from the energy of which an entity like an aggregator manages to gain profit [20]. Moreover, the penalty factor of CO_2 emissions is modeled in its main function, which shows the cost of GT. Figure 12.1 shows the correlation between VPP and other distributed generation resources.

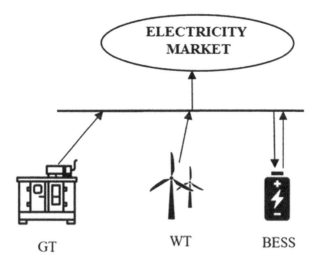

FIGURE 12.1   Structure of the VPP with its relationships.

In this chapter, the following assumptions have been made:

- VPP should act as a price taker player in power market; and
- Small BESS have been aggregated and converted into a large BESS to provide much more energy.

12.2.1 OBJECTIVE FUNCTION

The proposed VPP model contains several WTs, BESS, and GT. The objective of the VPP in the joint markets of energy is to maximize its obtained net profit as stated in Equation (12.1).

$$profit(h) = \sum_h \; \mathrm{Re}_E^{DA}(h) - Co^{DG}(h) \tag{12.1}$$

This objective function is composed of two different terms: the first term is the income of VPP earned by participating in the energy market, and the second term models a cost of the VPP from utilizing GT.

12.2.2 GT OPERATIONAL MODEL

The GTs are modeled as peak shaver generation and also used in critical situations. In other words, GTs are designed to fulfill the demanded load caused by other DPGs or at times of high prices to reduce the cost function [21]. However, due to the effects of GT activities on the climate change issue, different penalties are applied to limit pollutant emissions. Since CO_2 emissions related to electricity generation are one of the most harmful emissions [5, 22] this chapter employs the operational cost of the GT, as in Equation (12.2), in two different terms: the first term is the startup and generation cost, as in Equation (12.3), and the second term represents the CO_2 emission penalty cost, as in Equation (12.4). Equation (12.5) shows the ramp rate limits of the GT in each hour. Equation (12.6) mentions that the output power of the GT must be between the minimum and maximum bounds. In order to model the GT different mode logics, Equations (12.7) and (12.8) are considered. Equations (12.9) and (12.10) are implemented to ensure the minimum up and down time limits, respectively.

$$Co^{GT}(h) = Co_{P\&S}^{GT}(h) + Co_{Em}^{GT}(h) \tag{12.2}$$

$$Co_{P\&S}^{GT}(h) = a.P^{GT}(h)^2 + b.P^{GT}(h) + c.U^{GT}(h) + SUC \times U_{on}^{GT}(h) \tag{12.3}$$

$$Co_{Em}^{GT}(h) = P^{GT}(h).CO2^{GT}.\lambda^{co2} \tag{12.4}$$

$$RDN \leq P^{GT}(h) - P^{GT}(h-1) \leq RUP \tag{12.5}$$

$$P_{min}^{GT} \times U^{GT}(h) \leq P^{GT}(h) \leq P_{max}^{GT} \times U^{GT}(h) \tag{12.6}$$

$$U_{on}^{GT}(h) - U_{off}^{GT}(h) \leq U^{GT}(h) - U^{GT}(h-1) \tag{12.7}$$

$$U_{on}^{GT}(h) + U_{off}^{GT}(h) \leq 1 \tag{12.8}$$

$$[X^{on}(h-1) - H^{on}] \times [U^{GT}(h-1) - U^{GT}(h)] \geq 0 \tag{12.9}$$

$$[X^{off}(h-1) - H^{off}] \times [U^{GT}(h) - U^{GT}(h-1)] \geq 0 \tag{12.10}$$

12.2.3 WIND TURBINE OPERATION MODEL

This section applies the Weibull distribution function as a proper expression model to calculate output power of the WT during any forecast hour [1, 23] in which a relationship between WT power output $P^{WT}(h)$ and the wind speed $V^{WT}(h)$ is defined, implementing Equation (12.11). In this relation, V_{ci}^{WT} is the cut in speed, V_r^{WT} is the rated speed, V_{co}^{WT} is the cut-out speed, and P_r^{WT} is the rated power of the WT.

$$P^{WT}(h) = \begin{cases} 0 & 0 \leq V^{WT} \leq V_{ci}^{WT} \\ (\dfrac{V^{WT}(h) - V_{ci}^{WT}}{V_r^{WT} - V_{ci}^{WT}}) \times P_r^{WT} & V_{ci}^{WT} \leq V^{WT} \leq V_r^{WT} \\ P_r^{WT} & V_r^{WT} \leq V^{WT} \leq V_{co}^{WT} \\ 0 & V^{WT} \geq V_{co}^{WT} \end{cases} \tag{12.11}$$

12.2.4 Battery Energy Storage System Operational Model

Equation (12.12) calculates the available stored energy in BESS for each hour. Upper limits of BESS in charge and discharging operation are presented in Equations (12.13) and (12.14), respectively. Equation (12.15) stops contention between charge and discharge operating modes which determine to just one action. Minimum and maximum allowable energy stored in BESS is considered by Equation (12.16).

$$E^{BESS}(h) = E^{BESS}(h-1) + (P_{ch}^{BESS}(h) - P_{dch}^{BESS}(h)) \times \Delta h \quad \Delta h = 1 \quad (12.12)$$

$$P_{ch}^{BESS}(h) \le P_{ch,MAX}^{BESS}(h) \times U_{ch}^{BESS}(h) \quad (12.13)$$

$$P_{dch}^{BESS}(h) \le P_{dch,MAX}^{BESS}(h) \times U_{dch}^{BESS}(h) \quad (12.14)$$

$$U_{ch}^{BESS}(h) + U_{dch}^{BESS}(h) \le 1 \quad (12.15)$$

$$0 \le E^{BESS}(h) \le E_{MAX}^{BESS}(h) \quad (12.16)$$

12.2.5 Energy Trading Constraints

Energy balance constraint is one of the constraints that the VPP has to satisfy in its scheduling problem. The energy balance constraint is expressed by Equation (12.17), which states that the energy generated by the VPP components has to be equal to the energy consumed by the VPP activities. The VPP components include the GT, the BESS, and the WTs. The energy balance constraint ensures that the VPP does not produce more or less energy than it needs to meet its market obligations. The energy balance constraint also considers the efficiency of the BESS.

$$P^{GT}(h) + P^{WT}(h) + U_{dch}^{BESS}(h) \times Eff^{BESS} - \frac{U_{ch}^{BESS}(h)}{Eff^{BESS}} = P^{DA}(h) \quad (12.17)$$

General parameters related to the activities of GT including capacity, operational limits, fuel costs, CO_2 emissions (kg/MWh), and CO_2 emission costs ($/kg) are reported in Table 12.1 (adopted from [24, 25] with a few adjustments).

There are wind turbines in the test system. Figure 12.2 presents the average hourly WT generation.

Moreover, the VPP does not own a single large-scale BESS, but rather aggregates a large number of small residential BESS that are distributed among its customers. The aggregator entity gathers these small residential BESS and coordinates their operation to become impressive in the power market. The assumed BESS is a battery energy storage system that consists of 750 household batteries with a capacity of 4 kW that are connected to the VPP. The household batteries are distributed among the customers who participate in the VPP program. The household batteries can charge or discharge power from or to the grid according to the VPP control signals. The BESS

TABLE 12.1
GT characteristics

P_{max} (MW)	P_{min} (MW)	RUP (MW)	RDN (MW)	RU (h)	RD (h)	a ($/MW2)	b ($/MW)	c ($)	SUC ($)	CO_2^{GT} (kg/MWh)	$CO_2\lambda$ ($/kg)
3	0	2	2	1	1	0.0055	44.5	23.9	25	0.24	30

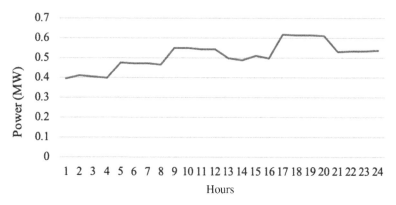

FIGURE 12.2   Hourly mean power generation by WT.

provides VPP with a total capacity of 3 MW, which can be used to balance the power supply and demand of VPP and increase its profit in the energy market. It has been assumed that the aggregated capacity of all residential BESS is equal to 3 MW, which means that the total amount of energy that can be stored by all residential BESS is 3 MWh. The minimum stored energy in the BESS should not be less than 300 kW. This assumption reflects the scale and potential of BESS in the VPP operation. BESS efficiency is assumed to be 95 percent, which means that 5 percent of the energy stored or discharged by BESS is lost due to conversion losses. The maximum charge and discharge rate of BESS is 200 kW per hour, which means that BESS cannot store or inject more than 200 kW of power in each hour. The proposed strategy is tested over a daily time period. Figure 12.3 shows forecast prices of energy market to sale provided energy.

Two different case studies are defined to illustrate the validity of the designed VPP model in the energy market. The case studies are based on different assumptions about the role of BESS in the VPP operation.

12.3   NUMERICAL STUDIES AND DISCUSSIONS

In this section, numerical studies and discussions are presented to evaluate the performance of the proposed VPP scheduling model. The VPP scheduling model is formulated as a mixed-integer linear programming (MILP) problem and solved in the GAMS environment. The objective of the VPP scheduling model is to maximize the daily profit of the VPP by optimally scheduling the power output of its components

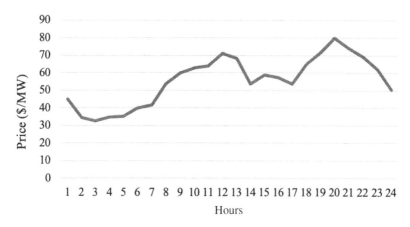

FIGURE 12.3 Day-ahead energy market prices.

and the energy sold in the electricity market. The VPP components include a GT, a BESS, and wind turbines. The parameters and data used in the numerical studies are based on realistic values from the literature and the electricity market. The numerical studies are conducted for two scenarios.

Scenario 1. The VPP trades in an energy market without taking into account the potential of BESS. In this case, the VPP relies only on the wind turbines and the gas turbine to supply power to the market.

Scenario 2. The VPP trades in an energy market considering the potential of BESS. In this case, the VPP utilizes the BESS to store or inject power to the market according to the market price and the state of charge of BESS.

As shown in Figure 12.4, the output power of the GT is a parameter that reflects the level of electric power that the turbine can produce by combusting fuel. The GT is determined by the operator based on the market price and the operational constraints of the turbine. The GT has been adjusted to a higher value from 8 a.m. to 11 p.m. in order to cope with the increase in energy market prices, which encourages the operator to utilize the turbine at a higher efficiency during those hours of high demand. The higher GT production also helps to meet the power requirement of VPP and increase its profit. In both scenarios, there is a positive correlation between the market prices and the GT productions, meaning that as the market price increased, the GT output power also increased.

Figure 12.5 depicts the load profile of BESS, which indicates the amount of power that BESS charges from or discharges to the network. The load profile of BESS is determined by the market prices and the state of charge of BESS. During the cheap hours (hours 1 to 7, 14, 16, and 17), BESS charges power from the network at a low cost and increases its state of charge. During the rest of the hours, when the market price is high, BESS discharges or injects power into the network and contributes to the power supply of VPP. Hence, using this strategy, BESS can optimize its operation and maximize the profit of VPP.

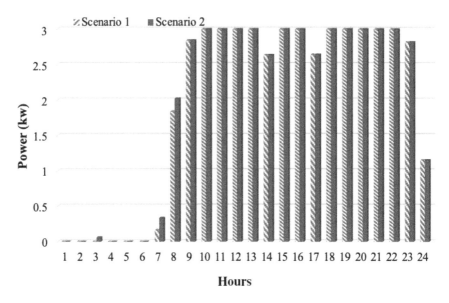

FIGURE 12.4   Output power of the gas turbine.

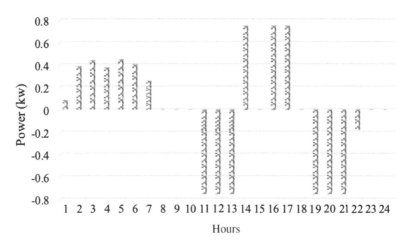

FIGURE 12.5   Charge and discharge profile of BESS.

Although the price of energy in hours 14, 16, and 17 is not lower than the price of energy in hours 1 to 7, BESS starts charging in these hours in order to prepare for injecting power into the network in the subsequent hours when the price of energy is higher. BESS uses a smart algorithm that considers the market price and the state of charge of BESS in different hours and compares them with each other to decide whether to charge or discharge. This is why BESS charges in hours 14, 16, and 17, even though the price of energy is not very low.

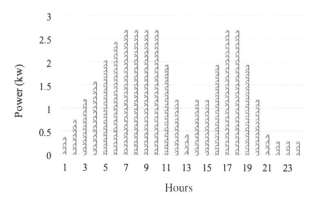

FIGURE 12.6 Hourly energy stored in BESS.

Figure 12.6 depicts the amount of energy stored in BESS in each hour of the day. The amount of energy stored in BESS is also known as the state of charge of BESS, which represents the percentage of the battery capacity that is filled with energy. The state of charge of BESS is determined by the charging and discharging power of BESS, which are controlled by the VPP operator based on the market price and the operational constraints of BESS. As is clear in the figure, BESS charges during hours 1–7, 14, 16, and 17, because the market prices are low during these hours and it is economical for the VPP to buy energy from the grid and store it in BESS. Hence, the state of charge of BESS are increased during these hours. During hours 11–13 and 19–22, in which VPP discharges the BESS because of high energy market prices, the state of charge of the BESS decreased. It should be noted that during hours 8–10, 15, 18, 23, and 24, VPP does not decide to charge or discharge the BESS. Hence, in these hours, the state of charge of the BESS remains constant. The figure shows that BESS operates intelligently and dynamically according to the market price and its state of charge.

Figure 12.7 illustrates the power that is sold to the electricity market by the VPP. This power is generated by a combination of WT, GT, and BESS. As can be seen in the figure, the power that is sold in the electricity market follows the same pattern as the market price, increasing from 8 to 23. Since the WT output power is uncertain and uncontrollable, this increase is mainly attributed to the GT and BESS, which can adjust their output according to the market conditions. In both scenarios, this trend implies that the power that is sold to the electricity market will increase as well, and consequently, the profit of the VPP will increase.

In Scenario 2, BESS is charged during the hours when the market price is low by utilizing the surplus power from the WT and the GT, and as a result, the energy that is sold to the electricity market (hours 1 to 7) is reduced. This enables BESS to save energy for future use when the market price is high. During hours 11–13 and 19–22 in which VPP discharges the BESS because of high energy market prices, the energy sold to the electricity market has been increased.

Figure 12.8 illustrates the hourly profit of VPP in a 24-hour time period. The profit of VPP is calculated as the difference between the revenue and the cost of VPP. The

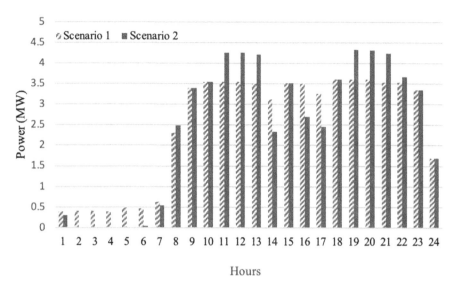

FIGURE 12.7   Power sold to the electricity market by VPP.

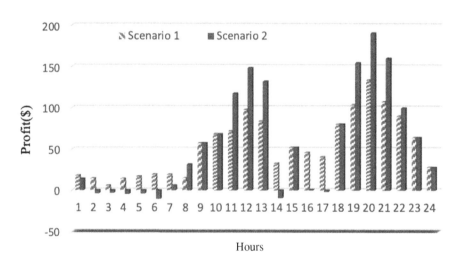

FIGURE 12.8   VPP hourly profit.

revenue of VPP is obtained from selling energy to the electricity market, while the cost of VPP is incurred from paying for the fuel of the gas turbine. The profit of VPP depends on the market price and the power output of VPP. With the increase in the electricity market price, the profit of VPP also increases because the energy sold to the electricity market by VPP also increases. This is especially true in the hours 9–13 and 18–23, when the market price is high and VPP utilizes its resources to generate more energy. On the other hand, in the hours 1–7, due to the low price of energy, the profit of the VPP decreased.

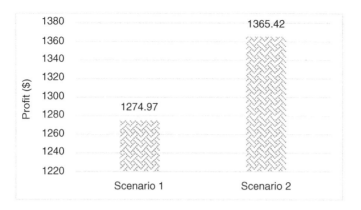

FIGURE 12.9 VPP daily profit for each scenario.

In Scenario 2, the VPP operates in an energy market considering the potential of BESS. In this scenario, the VPP uses BESS to store or inject power to the network according to the market price and the state of charge of BESS. In Scenario 2, the VPP during the hours 1–7, 14, 16, and 17 earned lower profits compared to Scenario 1. This is because the VPP charges BESS in these hours, when the market price is low or moderate, and consequently injects less power to the upstream network. The VPP expects to make up for this loss by discharging BESS in the following hours when the market price is higher. Hence, during hours 11–13 and 19–22, in which VPP discharges the BESS, more energy has been sold to the electricity market, and consequently more profit has been earned. The figure shows that VPP can optimize its profit by adjusting its power output according to the market price and its operational constraints.

Figure 12.9 compares the daily profit of VPP in two considered scenarios. The daily profit of VPP is the sum of the hourly profit of VPP in a 24-hour period. In Scenario 2, the VPP operates in an energy market considering the potential of BESS, while in Scenario 1, the VPP operates without considering the potential of BESS. As expected, in Scenario 2, due to the presence of BESS, the profit obtained for VPP is greater than in Scenario 1. This is because BESS enables the VPP to store energy when the market price is low and inject energy when the market price is high.

12.4 CONCLUSION

This chapter presents a VPP scheduling model that optimally operates its DPGs in a day-ahead electricity market. In the proposed model, the penalty cost of CO_2 emissions has also been considered to control the pollutant emissions of GT operation. The VPP clusters residential BESS by BESS aggregator and uses them to increase obtained profit. This chapter shows that the presence of BESS in VPP can be very effective for improving the economic performance of VPP and enhancing its flexibility and reliability in the electricity market. Numerical results demonstrate the efficiency and capability of the presented scheduling framework to guarantee the maximum expected profit achievement of the VPP.

NOMENCLATURE

Indices

H	Index of time intervals
L	Demand response programs type of energy index $l = 2,3,...,N_l$

Parameters

a, b, c	Gas turbine generation cost ($)
$CO2^{GT}$	Gas turbine CO_2 emissions (ton/kW)
H^{on}, H^{off}	Minimum up and down times of gas turbine (h)
$P_{min}^{GT}, P_{max}^{GT}$	Maximum and minimum limitation of gas turbine output power (kW)
$P_{dch,MAX}^{BESS}(h); P_{ch,MAX}^{BESS}(h)$	Maximum rate of charge and discharge of BESS in hour h (kW)
$P^{WT}(h)$	Power generated by wind turbines at hour h (kW)
P_r^{WT}	Nominal power of the wind turbines (kW)
RUP, RDN	Gas turbine ramp-up and ramp-down limitation (kW/h)
SUC	Gas turbines startup cost ($)
$V_{ci}^{WT}, V_r^{WT}, V_{co}^{WT}$	Cut in, rated, and cut-out speeds of the wind turbine (m/s)
$V^{WT}(h)$	Hourly wind speed (m/s)
λ^{co2}	CO_2 penalty price of emissions ($/ton)
$\lambda^E(h)$	Price of day-ahead energy market at hour h ($/kWh)

Variables

Continuous variables

$P^{DA}(h)$	Total energy sold in day-ahead market at hour h (kW)
$P^{GT}(h)$	Gas turbine power generation at hour h (kW)
$P_{ch}^{BESS}(h), P_{dch}^{BESS}(h)$	BESS power in charge and discharge modes at hour h (kW)
$RR^{GT}(h)$	GT regulation provision at hour h (kW)
$X^{on}(h), X^{off}(h)$	Time duration that gas turbine has been on and off at hour h

Binary variables

$U^{GT}(h)$	Binary decision variable for on/off state of GT at hour h (on=1, off=0)
$U_{on}^{GT}(h), U_{off}^{GT}(h)$	Binary decision variables for startup/shutdown state of GT at hour h
$U_{ch}^{BESS}(h), U_{dch}^{BESS}(h)$	Binary decision variables for on/off of BESS at hour h

REFERENCES

[1] M. Zhao, Z. Chen, F. Blaabjerg, "Probabilistic capacity of a grid connected wind farm based on optimization method," *Renew Energy*, vol. 31, p. 2171e2187, 2006.

[2] J. G. Wanjiku, M. A. Khan, P. S. Barendse and A. B. Sebitosi, "A. Analytical sizing of an electrolyser for a small scale wind electrolysis plant," *2010 IEEE International Energy Conference, Manama, Bahrain*, pp. 10–15, 2010.

[3] G. Koeppel, "Distributed generation literature review and outline of the Swiss situation," In EEH Power Systems Laboratory, Internal Report, Zurich, Germany, 2003.

[4] S. Gonçalves, H. Morais, T. Sousa and Z. Vale, "Energy resource scheduling in a real distribution network managed by several virtual power player," in *IEEE Power and Energy Society Transmission and Distribution*, Orlando, Florida, 2012.

[5] H. M. Ghadikolaei , E. Tajik, J. Aghaei, M. Charwand, "Integrated day-ahead and hour-ahead operation model of discos in retail electricity markets considering DGs and CO2 emission penalty cost," *Appl Energy*, vol. 95, p. 174–185, 2012.

[6] P. A. Awerbuch S, "The virtual utility: Accounting, technology & competitive aspects of the emerging industry," in *Kluwer Academic Publishers*, MA, USA, 1997.

[7] D. Pudjianto, C. Ramsay and G. Strbac, "Virtual power plant and system integration of distributed energy resources," *IET Gener Transm Distrib*, vol. 1, no. 1, pp. 10–16, 2007.

[8] J. Oyarzabal, J. Marti, A. Ilo, M. Sebastian, D. Alvira and K. Johansen, "Integration of DPG into power system operation through virtual power plant concept applied for voltage regulation," in *CIGRE/IEEE PES Joint Symposium*, pp. 1–7, 2009.

[9] D. Pudjianto, C. Ramsay and G. Strbac, " Microgrids and virtual power plants: Concepts to support the integration of distributed energy resources," in *IMechE Power and Energy*, vol. 222, pp. 731–741, 2008.

[10] Q. F. Y. L. a. J. C. Jie Yu, "Stochastic optimal dispatch of virtual power plant considering correlation of distributed generations," *Math Probl Eng*, vol. 2015, no. Article ID 135673, p. 8. http://dx.doi.org/10, 2015.

[11] S. Gonçalves, H. Morais, T. Sousa and Z. Vale, "Energy resource scheduling in a real smart grid managed by several virtual power player," *IEEE Power and Energy Society Transmission and Distribution*, vol. 2010, p. 7–10, 2012.

[12] C. Hutson, G. K. Venayagamoorthy and K. A. Corzine, "Intelligent scheduling of hybrid and electric vehicle storage capacity in a parking lot for profit maximization in grid power," in *IEEE Energy 2030*, pp. 1–8, 2008.

[13] O. Arslan, O. E. Karasan, "Cost and emission impacts of virtual power plant formation in plug-in hybrid electric vehicle penetrated networks," *Energy*, vol. 60, pp. 116–124, 2013.

[14] R. Raineri, S. Ríos and D. Schiele, "Technical and economic aspects of ancillary services markets in the electric power industry: an international comparison," *Energy Policy*, vol. 34, pp. 1540–1555, 2006.

[15] H. Pandžić, J. M. Morales, A. J. Conejo and I. Kuzle, "Offering model for a virtual power plant based on stochastic programming.," *Appl Energy*, vol. 105, pp. 282–292, 2013.

[16] A. Shayegan-Rad, A. Badri and A. Zangeneh, "A. Day-ahead scheduling for virtual power plant in joint energy and regulation reserve markets under uncertainties.," *Energy*, vol. 121, pp. 114–125, 2017.

[17] E. Mashhour and S. M. Moghaddas-Tafreshi, "Bidding strategy of virtual power plant for participating in energy and spinning reserve markets," *IEEE Trans Power Syst*, vol. 26, pp. 949–56, 2011.

[18]   A. Shayegan-Rad, A. Badri and A. Zangeneh, "Day-ahead scheduling for virtual power plant to participate in energy and spinning reserve markets," in *29th International Power System Conference (PSC 2014)*, Tehran, Iran, 2014.

[19]   S. S. Mahdavi and M. H. Javidi, "VPP decision making in power markets using Benders decomposition," *Int Trans Elect Energy Syst*, vol. 24, pp. 960–975, 2014.

[20]   A. Shayegan-Rad, A. Badri and A. Zangeneh, "A multi leader-follower game theory for optimal contract pricing of virtual power plants in smart distribution networks," *IET Gener Transm Distrib*, vol. 12, pp. 5747–5752, 2018.

[21]   A. Zakariazadeh, S. Jadid and P. Siano, "Economic-environmental energy and reserve scheduling of smart distribution systems: A multiobjective mathematical programming approach," *Energy Convers Manage*, vol. 78, pp. 151–164, 2014.

[22]   S. Wong, K. Bhattacharya and J. D. Fuller, "Long-term effects of feed-in tariffs and carbon taxes on distribution system," *IEEE Trans Power Del*, vol. 25, no. 3, pp. 1241–1253, 2010.

[23]   B. G., *Renewable energy*, Oxford University Press, Oxford, UK, 2004.

[24]   A. Shayegan-Rad, A. Badri and A. Zangeneh, "Smart microgrid energy and reserve scheduling with demand response using stochastic optimization," *Electr Power Syst*, vol. 63, pp. 523–33, 2014.

[25]   S. W. Hadley and J. W. Van Dyke, "Emissions benefit of distributed generation in the Texas market," Technical report 885960, prepared for the U.S. Department of Energy (DOE), 2003.

13 Role of Battery-Integrated Residential Systems for Regulation Reserve Markets

Ali Shayegan-Rad, Mohammad Reza Fallahzadeh, and Ali Zangeneh

13.1 INTRODUCTION

The small capacity and unpredictable nature of renewable energy sources can make it difficult for them to participate in energy and regulation reserve (RR) markets. To solve this problem, the concept of virtual power plant (VPP) is presented. The VPP acts as a new player in the power market, which is responsible for the coordination of small-scale distributed energy resources (DERs); these scattered generation units can be renewable and fossil fuel-based small power plants. By aggregating and coordinating small production units, the VPP creates an acceptable capacity to participate in the joint energy market. By integrating a large number of small distributed generation units, the VPP behaves like a powerful player in the energy market, which will be able to sell a large part of energy produced by the renewable sector in the electricity market, consequently helping global warming by reducing CO_2 emissions [1, 2].

The structure of a VPP is designed in such a way that distributed power generation sources can play a role in the distribution network. This issue is explained in detail in [3]. A VPP stochastic programming model that includes wind and gas turbines and integrates demand response programs is given in [4]. In [5] there is a general method for coordinating power generation sources that include battery storage equipment. This method deals with the maximum use of power generation resources and also enables scattered production resources to participate in the market in order to bring maximum economic efficiency. In [6], a technical and economic model for the regulation market is discussed. In this model, more is focused on the regulation market. A new model for scheduling pumped storage power plants in the rotating reserve market and regulation market is presented in [7]. This model will enable pumped storage power plants to participate in the day-ahead market and thus increase the efficiency of the electricity market. An operational model for modeling distributed generation resources under VPP that enables them to participate in the energy market and the rolling storage market is presented in [8, 9].

DOI: 10.1201/9781003441236-13

In [10], VPP has used a diverse number of production resources to optimize its profit in the energy market and the rotating storage market. The wide variety of power generation sources under the VPP's control allows the VPP to use the different characteristics of these sources to provide benefits. In [11], a VPP uses a planning process to utilize different power generation sources (which are randomly selected), including a gas turbine (GT) and a small storage pump power plant, which is used to participate in the previous-day market. The strength of this model of VPP is to maximize its profit through the use of various features of the resources under its control. In [12], a VPP is modeled in such a way that it uses sources with variable production power, such as wind turbine and solar cell, to participate in the energy market and RR. In [13], a three-step, two-level plan is stated in which the VPP makes proposals in the energy market and regulates the market. This structure enables the VPP to maximize its profit by making the right decisions about power supply in the market.

13.2 VPP MODELING

The proposed VPP structure includes the use of various distributed generation sources such as gas turbines, wind turbines, and batteries used in home systems. Also, in order to maximize the profit of the VPP, these resources have been used in both the energy and reserve markets. Figure 13.1 shows the structure of the proposed model and the relationship of different power generation sources with VPP.

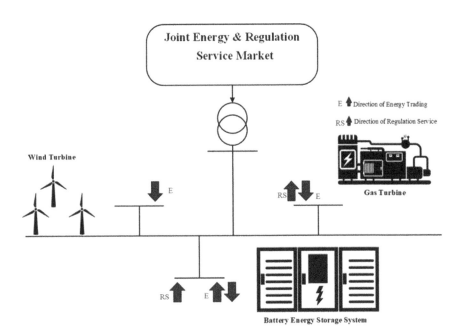

FIGURE 13.1   Foundation of the proposed virtual power plant.

13.3 VPP FORMULATION

The most important purpose of the VPP is to maximize its total profit by participating in both the energy and RR markets proactively. This is attainable through a mathematical formation represented by Equation (13.1), which is formed by the difference between incomes and cost.

$$profit(h) = \sum_h \text{Re}_E^{DA}(h) + \text{Re}_{RR}^{DA}(h) - Co^{GT}(h) \tag{13.1}$$

This equation is composed of three terms: the first and second terms express income obtained from participation in the energy and RR markets, while the third component is designed to show the whole cost that the VPP has to pay associated with the utilization of the gas turbine (GT). This term, which is known as cost of VPP, directly affects the overall objective function of the VPP. Therefore, by reducing the GT cost and increasing the revenue through actively participating in the power and RR market, VPP profit is maximized.

Equation (13.2) describes the third component of the objective function of the VPP, which consists of different elements. In this equation, the first parameter is the cost of generation of electricity by gas turbine, while the second term is the penalty factor for CO_2 emissions. Equation (13.2) is expressed in detail by Equations (13.3) and (13.4), respectively. Equation (13.3) relates to the different characteristics of the gas turbine. Equation (13.4) shows the penalty factor of GT in each hour. Equation (13.5) demonstrates the limitation of the GT in ramp-up and ramp-down. These constraints are vital in maintaining the stability and efficiency of the GT's operation. Equation (13.6) shows the acceptable upper and lower limits of GT generation; this ensures that the GT operates within its designed capacity.

Equations (13.7) and (13.8) are used to stop conflict within the VPP planning, which is essential to ensure the coordinated functioning of various components within the VPP. Equations (13.9) and (13.10) guarantee minimum up and down time limits, respectively. Finally, Equations (13.11) and (13.12) define the upper and lower limitations of the RR provision. This ensures that the VPP is able to meet its reserve obligations while operating within its capacity limits. In summary, these boundaries ensure that the VPP remains in a stable situation while maximizing profit.

$$Co^{GT}(h) = Co_G^{GT}(h) + Co_{Em}^{GT}(h) \tag{13.2}$$

$$Co_G^{GT}(h) = a.P^{GT}(h)^2 + b.P^{GT}(h) + c.U^{GT}(h) + SUC \times U_{on}^{GT}(h) \tag{13.3}$$

$$Co_{Em}^{GT}(h) = P^{GT}(h).CO2^{GT}.\lambda^{co2} \tag{13.4}$$

$$RDN \leq P^{GT}(h) - P^{GT}(h-1) \leq RUP \tag{13.5}$$

$$P_{min}^{GT} \times U^{GT}(h) \leq P^{GT}(h) \leq P_{max}^{GT} \times U^{GT}(h) \tag{13.6}$$

$$U^{GT}_{on}(h) - U^{GT}_{off}(h) \le U^{GT}(h) - U^{GT}(h-1) \tag{13.7}$$

$$U^{GT}_{on}(h) + U^{GT}_{off}(h) \le 1 \tag{13.8}$$

$$[X^{on}(h-1) - H^{on}] \times [U^{GT}(h-1) - U^{GT}(h)] \ge 0 \tag{13.9}$$

$$[X^{off}(h-1) - H^{off}] \times [U^{GT}(h) - U^{GT}(h-1)] \ge 0 \tag{13.10}$$

$$RR^{GT}(h) + P^{GT}(h) \le P^{GT}_{max}.U^{GT}_{on}(h) \tag{13.11}$$

$$RR^{GT}(h) \le P^{GT}(h) \tag{13.12}$$

The Weibull distribution function is used to calculate WT generations [8]. Equation (13.13) indicates the output power of the wind turbine $P^{WT}(h)$ using related wind speed $V^{WT}(h)$.

$$P^{WT}(h) = \begin{cases} 0 & 0 \le V^{WT} \le V^{WT}_{ci} \\ (\dfrac{V^{WT}(h) - V^{WT}_{ci}}{V^{WT}_{r} - V^{WT}_{ci}}) \times P^{WT}_{r} & V^{WT}_{ci} \le V^{WT} \le V^{WT}_{r} \\ P^{WT}_{r} & V^{WT}_{r} \le V^{WT} \le V^{WT}_{co} \\ 0 & V^{WT} \ge V^{WT}_{co} \end{cases} \tag{13.13}$$

Equation (13.14) is used to calculate the total energy stored in the BIRS in each hour, which takes into account the RR delivery request for participation in the regulation market. This is the most important equation determining the BIRS's capacity to meet reserve requirements while participating in the market. Equations (13.15) and (13.16) define the upper limits of the BIRS in charge and discharge modes, respectively. To avoid conflict between charge and discharge modes, Equation (13.17) is implemented. Equation (13.18) shows the minimum and maximum capacity of the BIRS. Finally, Equations (13.19) and (13.20) apply permissible regulation up and down, respectively.

$$\begin{aligned} E^{BIRS}(h) = E^{BIRS}(h-1) + (P^{BIRS}_{ch}(h) - P^{BIRS}_{dch}(h) - (\text{Pr}^{reg}_{up} - \text{Pr}^{reg}_{Down}) \\ \times RR^{DA}(h)) \times \Delta h \quad \Delta h = 1 \end{aligned} \tag{13.14}$$

$$P^{BIRS}_{ch}(h) \le P^{BIRS}_{ch,MAX}(h) \times U^{BIRS}_{ch}(h) \tag{13.15}$$

$$P^{BIRS}_{dch}(h) \le P^{BIRS}_{dch,MAX}(h) \times U^{BIRS}_{dch}(h) \tag{13.16}$$

$$U^{BIRS}_{ch}(h) + U^{BIRS}_{dch}(h) \le 1 \tag{13.17}$$

$$0 \le E^{BIRS}(h) \le E^{BIRS}_{MAX}(h) \tag{13.18}$$

$$RR^{BIRS}(h) + P_{dch}^{BIRS}(h) \le P_{dch.MAX}^{BIRS}(h) \times U_{dch}^{BIRS}(h) \qquad (13.19)$$

$$RR^{BIRS}(h) \le P_{dch}^{BIRS}(h) \qquad (13.20)$$

Equation (13.21) demonstrates the revenue that the VPP earns from its participation in the energy market. This equation is vital because it affects the objective function of the VPP directly, as well as the amount the VPP gains from selling energy into the grid. Another revenue that VPP can obtain in this model is generated from providing regulation services. This revenue, outlined in Equation (13.22), consists of two terms: the first term represents the income that the VPP earns from declaring its readiness to provide regulation services, while the second term shows the revenue or cost related to delivering up/down regulation as requested, based on spot prices. According to Equation (13.23), spot market prices are determined using random parameters p1 and p2, as referenced in [8]. This equation models the inherent uncertainty and variability in energy market prices. These random parameters capture a range of factors that can influence spot prices, including demand and supply conditions, weather patterns, and grid stability requirements.

$$\text{Re}_E^{DA}(h) = P^{DA}(h) \times \lambda^E(h) \qquad (13.21)$$

$$\text{Re}_{RR}^{DA}(h) = RR^{DA}(h) \times \lambda^{RR}(h) + (\text{Pr}_{up}^{reg}(h) - \text{Pr}_{down}^{reg}(h)) \times RS^{DA}(h) \times \lambda^{spot}(h) \qquad (13.22)$$

$$\lambda^{spot}(h) = \begin{cases} (1+p_1) \times \lambda^E(h) & 0 \le p_1 \le 0.2 \quad h \in [12-23] \\ (1+p_2) \times \lambda^E(h) & -0.1 \le p_2 \le 0.1 \quad otherwise \end{cases} \qquad (13.23)$$

Equation (13.24) is facilitated to ensure a balance between the energy produced by the VPP and the energy sold in the electricity market. Equation (13.25) is implemented to maintain energy balance between the RR provided by the VPP and what is sold to the electricity market. They ensure a balance between production and sales, in terms of both energy and regulatory services. By achieving these balances, the VPP can operate stably and maximize profit while fulfilling its duty in the electricity market.

$$P^{GT}(h) + P^{WT}(h) + P_{DCh}^{BIRS}(h) = P^{DA}(h) + \frac{P_{Ch}^{BIRS}(h)}{\eta} + (\text{Pr}_{up}^{reg}(h) - \text{Pr}_{down}^{reg}(h)) \times RR^{DA}(h)$$

$$(13.24)$$

$$RR^{GT}(h) + RR^{BIRS}(h) = RR^{DA}(h) \qquad (13.25)$$

13.4 NUMERIC STUDIES AND DISCUSSIONS

The strategy proposed in this study is evaluated over a 24-hour period. Figure 13.2 provides a visual representation of the forecast prices in the energy market, as referenced in [14].

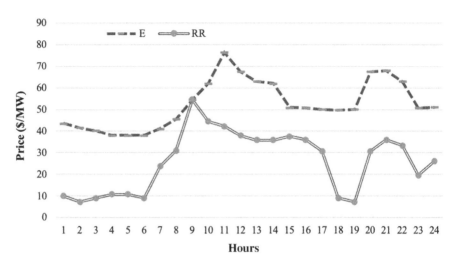

FIGURE 13.2 Day-ahead energy and RR market prices.

TABLE 13.1
GT characteristics

P_{max} (MW)	P_{min} (MW)	RUP (MW)	RDN (MW)	RU (h)	RD (h)	a ($/MW)	b ($/MW)	c ($)	SUC ($)	CO_2^{GT} (kg/MWh)	$CO_2\lambda$ ($/kg)
5	0	2	2	1	1	0.0055	44.5	23.9	25	0.78	75

Table 13.1 presents a comprehensive list of general parameters related to the operations of the GT. These parameters include capacity, operational limits, fuel costs, CO_2 emissions (kg/MWh), and CO_2 emission costs ($/kg). These parameters provide a detailed overview of the various factors that influence the operations and performance of the GT. The capacity and operational limits define the scale and scope of the GT's activities. The fuel costs and CO_2 emission costs provide insights into the financial implications of the GT's operations. The CO_2 emissions parameter, expressed in kg/MWh, quantifies the environmental impact of the GT's activities. The data presented in Table 13.1 are adopted from [15, 16], with a few adjustments made to fit the specific context of this study. This ensures that the data are not only accurate and reliable but also relevant to the current analysis.

In this scenario, the VPP does not possess a single large-scale battery-integrated residential system (BIRS), but rather it consolidates numerous small residential BIRS units that are distributed among its customer base. This aggregation is managed by an entity that coordinates the operation of these small residential BIRS units, enabling them to have a significant impact in the power market. It is assumed in this chapter that the BIRS is a combination of 2000 household batteries, each with a capacity of 4 kW, and all connected to VPP. They are able to charge or discharge to the network based on the signal received from VPP. The total capacity of the BIRS aggregated from 2000 household batteries is 8 MW. This is a great capacity to help balance energy supply and demand based on VPP. Also, by using this capacity and helping to improve the technical characteristics of the network, it is possible to generate revenue by means of the VPP. It is assumed that the combined capacity of all residential BIRS units equals 8 MW, translating to a total energy storage potential of 8 MWh. The minimum energy stored in the BIRS should not fall below 800 kW, reflecting the scale and potential of the BIRS in VPP operations.

The efficiency of the BIRS converter is assumed to be 95 percent, indicating that 5 percent of the energy stored or discharged by the BIRS is lost due to conversion losses. The maximum charge and discharge rate of the BIRS is set at 1 MW per hour, meaning that it cannot store or inject more than 1 MW of power within an hour. Furthermore, incentive contracts are considered for encouraging residential consumers to participate in RR provision.

The test system includes WTs as a part of its energy generation infrastructure. Figure 13.3 shows the average hourly generation output of these WTs. The generation output of WTs can vary significantly based on factors such as wind speed, turbine design, and operational efficiency.

Figure 13.4 illustrates the requests for regulation up and down services. These up and down requests manage the balance between supply and demand in real time, ensuring that the frequency and voltage of the grid remain within safe and operational limits. Regulation up refers to the ability to increase power output or decrease power consumption in response to a rise in demand or a drop in supply. Conversely, regulation down involves reducing power output or increasing power consumption when there is excess supply or reduced demand.

In essence, Figure 13.4 serves as a comprehensive visual representation of regulation up and down requests, highlighting their importance in grid operations and the role of the VPP in fulfilling these requests.

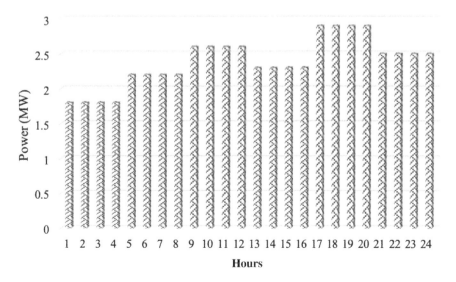

FIGURE 13.3   Hourly mean power generation of WT.

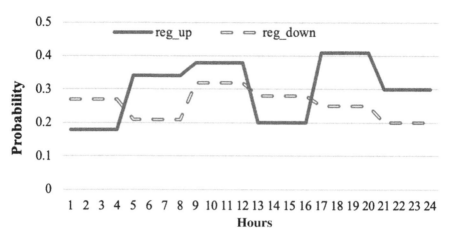

FIGURE 13.4 Regulation up and down calling requests.

Two different case studies are presented to illustrate of the proposed VPP scheduling model in joint markets of energy and RR:

Scenario 1. The VPP trades in an energy and RR markets without considering the potential of BIRS.

Scenario 2. The VPP trades in an energy and RR markets considering the potential of BIRS.

Figure 13.5 illustrates a gas turbine output power. It is evident from the figure that when energy prices are high, there is a corresponding increase in the output power of the GT. This suggests a direct relationship between energy costs and the performance of the GT.

Figure 13.6 presents the RR provided by the GT. It is efficient that a significant portion of the power generated by the GT has been bid into the RR market. As shown in the figure, the VPP decides to procure the maximum RR which is equal to provided energy.

As depicted in Figure 13.7, the BIRS strategically charges and discharges based on the fluctuation of energy prices during peak and non-peak hours. This strategy is implemented with the aim of maximizing profit.

During periods of high energy demand (peak hours), when energy prices are typically higher, the BIRS discharges stored energy. Conversely, during periods of low demand (non-peak hours), when energy prices are lower, the BIRS charges or stores energy. This cycle of charging and discharging allows the BIRS to buy energy at a lower cost and sell it at a higher price, thereby maximizing profit. In essence, Figure 13.7 illustrates this strategic operation of BIRS, providing a clear understanding of how energy storage systems can be leveraged for economic benefits in the energy market.

Figure 13.8 provides a depiction of the energy stored in the BIRS. This figure shows that the amount of stored energy in the BIRS is inversely proportional to the

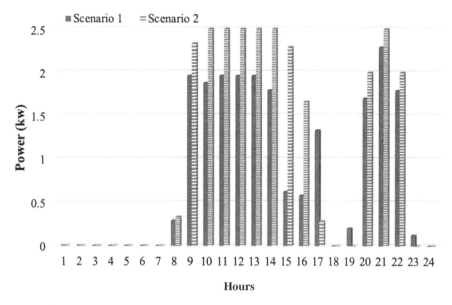

FIGURE 13.5   Hourly generated power by gas turbine.

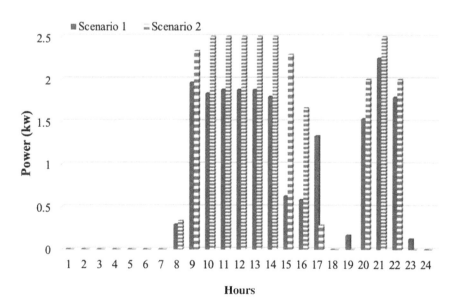

FIGURE 13.6   Regulation reserve bid by gas turbine.

price of energy. Specifically, when energy prices are high, the BIRS tends to store less energy. This is likely because it is more profitable to sell or discharge the stored energy during these periods of high prices. On the other hand, when energy prices are

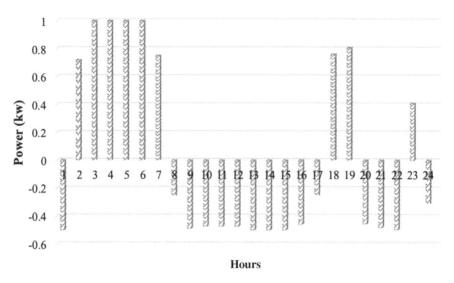

FIGURE 13.7   Charge and discharge of BIRS during day and night.

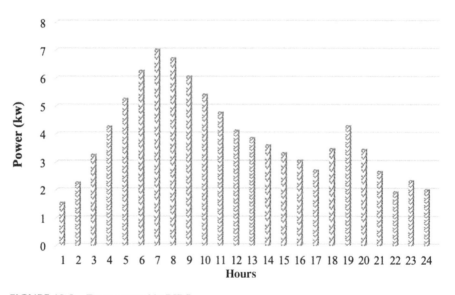

FIGURE 13.8   Energy stored in BIRS.

low, the BIRS stores more energy, taking advantage of the lower cost to maximize its energy reserves.

Figure 13.9 shows the RR bid by the BIRS. It is important that during non-peak hours, the BIRS does not make any offers. This is likely due to the absence of calling requests during these periods. This suggests that the BIRS strategically participates in the RR market, aligning its bidding behavior with market demand. During non-peak hours, when demand is low and there are no calling requests, the BIRS refrains

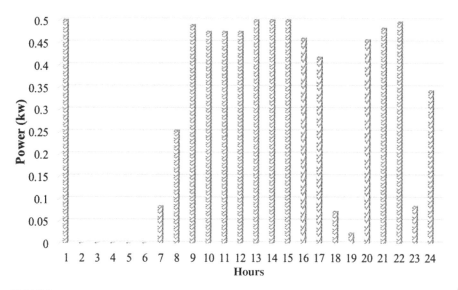

FIGURE 13.9 Regulation reserve bid by BIRS.

from bidding. This approach could potentially optimize the use of stored energy and enhance the profitability of BIRS.

Figure 13.10 presents a comparison of the hourly energy sold to the market by the GT, BIRS, and WT under two different scenarios. The comparison reveals that when the BIRS is integrated into the system, the amount of energy sold to the electricity market in peak hours is significantly higher than when the BIRS is absent.

In other words, in Scenario 2, during non-peak hours, the amount of energy sold to the electricity market is less when the BIRS is present due to the policy that the BIRS stores energy during non-peak hours. During non-peak hours in Scenario 2, the amount of energy sold to the electricity market is lower when the BIRS is present. This is because of the policy that the BIRS stores energy during non-peak hours. As a result, there is less energy available for sale during non-peak hours. However, during peak hours, the BIRS can act as a generation unit to feed additional demand, which can help the growth of demand in future and not restructuring the power grid. Also, this feature helps to postpone the renovation of the power plants, making it unnecessary to build new plants and reducing the need to use expensive plants in peak hours. Therefore, while the BIRS facilitates the power market, it results in a reduction in energy sales during non-peak hours, and it can provide substantial advantages such as technically and commercially during peak hours.

Figure 13.11 illustrates the aggregated RR bid to the electricity market by GT and BIRS, considering both up and down regulation calling requests. The figure shows that when the BIRS is integrated into the formulation, there is an outstanding increase in the amount of RR provision. This suggests that adding the capacity of the BIRS into the energy system results in a notable increase in the market participation of power generators, which is conducive to an increased level of power bids for RR. This

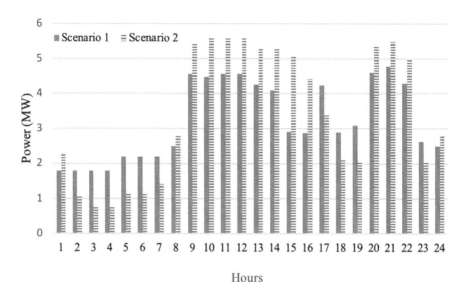

FIGURE 13.10   Electricity sold to the electricity market.

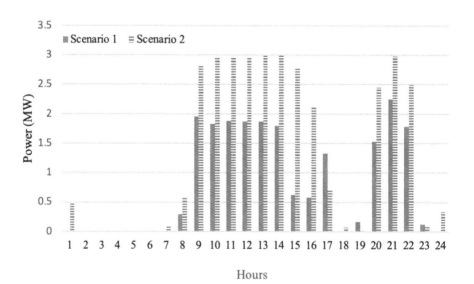

FIGURE 13.11   RR sold to the electricity market.

feature enhances the dynamics and reliability of the power market, which is better able to handle fluctuations in demand.

Figure 13.12 provides a comparison of the profits obtained over a 24-hour period, with and without the integration of the BIRS. It is evident from the figure that when the BIRS is incorporated into the system, the parameters of the electricity market

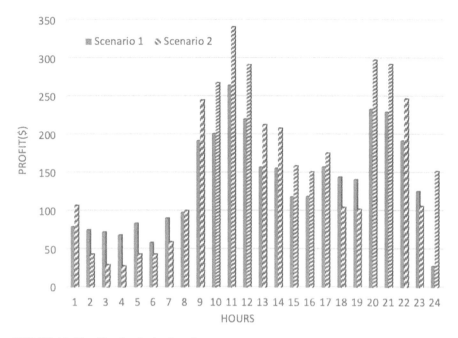

FIGURE 13.12   Hourly obtained profit.

improve significantly. As a result, the scenario with BIRS proves to be more profitable during peak hours compared to the scenario without BIRS. When the BIRS is charged during off-peak hours, less power is sold to the electricity market, which results in lower profits. This shows that the BIRS acts as a crucial player in improving the profitability of the VPP. Integrating the BIRS into the power system results in a more efficient and competitive electricity market. The BIRS also contributes to a more reliable distribution network by improving the technical characteristics of the network.

Figure 13.13 presents a comparison of the total profits accrued over a 24-hour period under two different scenarios. The figure clearly demonstrates that the scenario with BIRS results in higher total profits over the course of 24 hours compared to the scenario without BIRS. This suggests that the inclusion of the BIRS in the energy system can significantly enhance its profitability.

13.5   CONCLUSION

This chapter presents a VPP programing framework that optimally operates its integrated resources in a day-ahead joint energy and RR market. In the presented framework, the incurred cost of CO_2 pollution has also been considered to decrease the pollutant emissions of the GT. Also, the VPP uses the BIRS in both energy and regulation modes. To demonstrate the robustness of the framework, two different scenarios have been applied to investigate the role of the BIRS in the VPP's expected profit. The results show that using the BIRS increased the profits generated by the

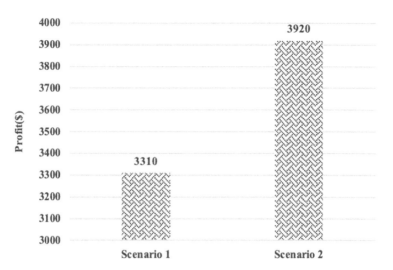

FIGURE 13.13   Total profit.

VPP. The numerical results illustrate the efficiency of the presented programing model for obtaining the maximum expected profit.

NOMENCLATURE

Indices

H	Index of time intervals
L	Index of steps in energy type of demand response programs $l = 2, 3, ..., N_l$

Parameters

a, b, c	Production cost coefficients of gas turbine ($)
$CO2^{GT}$	CO_2 emission rate of gas turbine (ton/kW)
H^{on}, H^{off}	Minimum up and down times of gas turbine (h)
$P^{GT}_{min}, P^{GT}_{max}$	Minimum and maximum limits of gas turbine output power (kW)
$P^{BIRS}_{dch,MAX}(h); P^{BIRS}_{ch,MAX}(h)$	Maximum limits of BIRS output power in charge and discharge modes (kW)
$P^{WT}(h)$	Power generation of wind turbines at hour h (kW)
P^{WT}_r	Rated power of the wind turbines (kW)
$Del^{reg}_{up}(h), Del^{reg}_{down}(h)$	Amount of up/down regulation calling requests at hour h
RUP, RDN	Ramp-up and ramp-down rate limits of gas turbine (kW/h)
SUC	Gas turbine startup cost ($)

$V_{ci}^{WT}, V_r^{WT}, V_{co}^{WT}$	Cut in, rated and cut-out speeds of the wind turbine (m/s)
$V^{WT}(h)$	Hourly wind speed (m/s)
λ^{co2}	CO_2 penalty price of emissions (\$/ton)
$\lambda^E(h)$	Price of day-ahead energy market at hour h (\$/kWh)
$\lambda^{RR}(h)$	Price of day-ahead regulation reserve market at hour h (\$/kWh)
$\lambda^{spot}(h)$	Price of spot market at hour h (\$/kWh)

VARIABLES

Continuous variables

$P^{DA}(h)$	Total energy sold in day-ahead market at hour h (kW)
$P^{GT}(h)$	Gas turbine power generation at hour h (kW)
$P_{ch}^{BIRS}(h), P_{dch}^{BIRS}(h)$	BIRS power in charge and discharge modes at hour h (kW)
$RR^{DA}(h)$	Total provided regulation reserve for day-ahead market at hour h (kW)
$RR^{BIRS}(h)$	Regulation reserve provided BIRS at hour h (kW)
$RR^{GT}(h)$	GT regulation provision at hour h (kW)
$X^{on}(h), X^{off}(h)$	Time duration that gas turbine has been on and off at hour h

Binary variables

$U^{GT}(h)$	Binary decision variable for on/off state of GT at hour h (on= 1, off=0)
$U_{on}^{GT}(h), U_{off}^{GT}(h)$	Binary decision variables for startup/shut down state of GT at hour h
$U_{ch}^{BIRS}(h), U_{dch}^{BIRS}(h)$	Binary decision variables for on/off of BIRS at hour h

REFERENCES

[1] ENIRDGnet: Concepts and opportunities of Distributed Gneration. "The Driving European Forces and Trend", ENIRDGnet Project deliverable D3, 2003.

[2] C Ziogou, D Ipsakis, P Seferlis, S Bezergianni, S Papadopoulou, S Voutetakis. "Optimal production of renewable hydrogen based on an efficient energy management strategy", *Energy* 2013;55:58–67.

[3] A Zangeneh, A Shayegan-Rad, F Nazari. "Multi-leader–follower game theory for modelling interaction between virtual power plants and distribution company", *IET Gener. Transm. Distrib.* 2018;12(21):5747–5752.

[4] A Shayegan-Rad, A Badri, A Zangeneh. "Day-ahead scheduling for virtual power plant in joint energy and regulation reserve markets under uncertainties", *Energy* 2017;121:114–125.

[5] F Bignucolo, R Caldon, V Prandoni, S Spelta, M Vezzola.. "The voltage control on MV distribution networks with aggregated DG units (VPP)", *Proceedings of the 41st*

International Universities Power Engineering Conference, Newcastle upon Tyne, UK 2006;1:187–192. doi: 10.1109/UPEC.2006.367741.

[6]   H Mohsenian-Rad. "Optimal demand bidding for time-shiftable loads", *IEEE Trans. Power Syst.* 2015;30(2):939–951.

[7]   S Nojavan, HA Aalami. "Stochastic energy procurement of large electricity consumer considering photovoltaic, wind turbine, micro-turbines, energy storage system in the presence of demand response program", *Energy Convers. Manage.* 2015;103:1008–1018.

[8]   A Shayegan-Rad, A Badri, A Zangeneh, M Kaltschmitt. "Risk-based optimal energy management of virtual power plant with uncertainties considering responsive loads", *Int. J. Energy Res.* 2019; 43:1–16.

[9]   J Qiu, K Meng, Y Zheng, ZY Dong. "Optimal scheduling of distributed energy resources as a virtual power plant in a transactive energy framework", *IET Gen. Trans. Dist.* 2017;11(13):3417–3427.

[10] SJ Kazempour, M ParsaMoghaddam, MR Haghifam, GR Yousefi. "Risk constrained dynamic self-scheduling of a pumpedstorage plant in the energy and ancillary service markets", *Energy Conver. Manage.* 2009;50(5):1368–1375.

[11] G He, Q Chen, C Kang, P Pinson, Q Xia. "Optimal Bidding Strategy of Battery Storage in Power Markets Considering Performance-Based Regulation and Battery Cycle Life," in *IEEE Transactions on Smart Grid*, 2016; 7(5): 2359-2367. doi: 10.1109/TSG.2015.2424314.

[12] S Papathanassiou, N Hatziargyriou, K Strunz. "A benchmark low voltage microgrid network", In: Proc CIGRE Symp Power SystDispersGenerat, Athens, Greece, April 17–20; 2005.

[13] E G Kardakos, C K Simoglou, A G Bakirtzis. "Optimal offering strategy of a virtual power plant: A stochastic Bi-Level approach", *IEEE Trans. Smart Grid* 2016; 7(2): 794e806.

[14] M Zhao, Z Chen, F Blaabjerg. "Probabilistic capacity of a grid connected wind farm based on optimization method" *Renew. Energy* 2006;31:2171e2187.

[15] A Zakariazadeh , S Jadid, P Siano. "Smart microgrid energy and reserve scheduling with demand response using stochastic optimization" *Electr. Power Syst.* 2014;63:523–33.

[16] S W Hadley, J W Van Dyke. "Emission benefit of distributed generation in the Texas market", U.S. Department of Energy (DOE), (Publication Number ORNL/TM-2003/100), 2003, Oak Ridge, Tennessee. Retrieved from : https://info.ornl.gov/sites/publications/Files/Pub57466.pdf

Index

Note: Page numbers in **bold** refer to tables and *italics* refer to figures.

For Product Safety Concerns and Information please contact our EU
representative GPSR@taylorandfrancis.com Taylor & Francis Verlag GmbH,
Kaufingerstraße 24, 80331 München, Germany

Printed and bound by CPI Group (UK) Ltd, Croydon, CR0 4YY
01/05/2025
01858556-0002